Polymer Science Library 3

POLYVINYL CHLORIDE DEGRADATION

Polymer Science Library

Edited by A.D. Jenkins
University of Sussex,
The School of Molecular Sciences,
Falmer, Brighton BN1 9QJ, England

Polymer Science Library 3

POLYVINYL CHLORIDE DEGRADATION

Jerzy Wypych

University of Ife, Ile-Ife, Nigeria

ELSEVIER

Amsterdam – Oxford – New York – Tokyo   1985

ELSEVIER SCIENCE PUBLISHERS B.V.
Sara Burgerhartstraat 25
P.O. Box 211, 1000 AE Amsterdam, The Netherlands

Distributors for the United States and Canada

ELSEVIER SCIENCE PUBLISHING COMPANY INC.
52, Vanderbilt Avenue
New York, N.Y. 10017

Library of Congress Cataloging-in-Publication Data

Wypych, Jerzy.
   Polyvinyl chloride degradation.

   (Polymer science library ; 3)
   Bibliography: p.
   Includes index.
   1. Polyvinyl chloride-Deterioration. I. Title.
II. Series: Polymer science library ; v. 3.
TP1180.V48W97 1985   668.4'237   85-16325
ISBN 0-444-42549-7 (U.S.)

ISBN 0-444-42549-7 (Vol. 3)
ISBN 0-444-41832-6 (Series)

Printed in The Netherlands

PREFACE

Creativity in research is based on past experience, objective evaluation of results and a hope of solving existing problems in future works. Previous research must always be critically scrutinized. Present understanding is full of confusion and question marks, but the future carries hope and mystery, and that is why it attracts the attention of new investigators.

This monograph mainly attempts to present recent efforts behind our current understanding, but controversial opinions are also discussed in order to provide material for future research. It seems that the most important area for detailed study is that of diffusion-controlled degradative processes, which should be closely related to the reaction environment. Also, more attention should be given to the presentation of studies in the form of functional relationships that possess more than purely descriptive value. Finally, one hopes that this data will provide a basis for mathematical modelling, which is so important for the practical application of polymers, especially in the context of stabilization.

Jerzy Wypych

Ile-Ife, Nigeria
27/06/85

TABLE OF CONTENTS

INTRODUCTION

The development of any field of technology or science is never independent of parallel contributions in other related areas. After more than 50 years of experimentation in the field of PVC science and technology, it is important to analyze the chronology of events. The data for the discussion are collected in Table 1.

TABLE 1.

The chronology of events in PVC science and technology (1).

Year	The most important achievements	Ref.
1797:	Extruder design	2
1835:	VC synthesis	3
1836:	Construction of calender	4
1854:	VC structure confirmed	5
1857:	Design of 3-roll mill	6
1865:	First plastification of polymer	7
1872:	Injection molding use	8
	Accidental VC polymerization	9
1894:	Industrial production of carbide and acetylene	10
1909:	Idea of emulsion and suspension polymerization	11,12
1912:	Industrial process of VC synthesis	13
1913:	Suggestion on possibility of PVC plastification	14
	Patents on PVC production and processing in form of organosols	14,15
	Patent on application of phosphoric acid esters for fire-proof products	14
1918:	VC thermal polymerization	16

1920:	Application of phthalates as PVC plasticizers	17,18
1922:	Studies on kinetics of VC polymerization	19
1926:	Studies on PVC properties (molecular weight, elementary analysis)	20
1928:	Application of prestabilizers	21
1930:	PVC defined as a mixture of compounds of varying chain length	22
	Thermal polymerization of VC with partial monomer conversion to obtain molecular weight uniformity	23
	"Head-to-tail" and "head-to-head" structures predicted	22
	Observations of PVC degradation during dissolution in hot solvents	22
1931:	VC emulsion polymerization	24
	PVC chlorination process	25
1932:	The use of enameled equipment	26
1933:	First suspension polymerization (methyl methacrylate)	27
1934:	First attempts at stabilization	28
	Alkaline polymerization	26
	Lead stabilizers applied	29
	Cadmium soap stabilizers applied	30
1935:	Elimination of low molecular weight polymer from polymerizate	31
	VC suspension polymerization	32
1937:	Spray drying	33
	Polymerization under inert gas	34
1938:	Sebacic esters as PVC plasticizers	35,36
	The mechanism of PVC emulsion polymerization	

From the table above, one can see that the top of the list contains achievements in the design of equipment for the processing of polymers. These machines were already well elaborated in the eighteenth and nineteenth centuries, long before chemistry was ready to yield any synthetic polymer. For the beginning of research on PVC we can cite the year 1835 in which Regnault (3) synthesized vinyl chloride. Some time was still needed until, in 1854, Kolbe (5) was able to define its chemical structure and to propose the name we use today. Other experiments with vinyl chloride were still under way, and as a consequence of its properties, they resulted in the first synthesis of polymer in 1872 by Baumann (9). One cannot be certain today if Baumann was exactly aware of the facts he observed, but his experiment was precisely described and ready to be repeated.

By the end of the nineteenth century, chemistry was again successful when the industrial production of carbide and acetylene was initiated. Chemists, happy to be able to supply the new convenience, developed carbide production very rapidly. Unfortunately for investors, the use of the acetylene flame did not last very long; electricity was soon to come. Therefore, chemists were forced to look at all possible ways of finding an outlet for large carbide production, in order to channel it to other effective areas.

It was quite fortunate that the German chemist, Klatte, who was not only a good professional but also something of a visionary, happened to be working around that time and resolved most of the problems connected with PVC production and processing (13-15). In consecutive patents he demonstrated how to produce vinyl chloride on an industrial scale, how to make polymer out of it, and how to process it in the form of organosols and plastisols using plasticizers for ordinary and self-extinguishing products. Unfortunately, he came too early with his ideas, and nobody tried to understand what he said, nor did anyone attempt to apply it directly. The situation in the chemistry field was just too difficult around that time, as nobody was really able to comprehend the basic principles of the process.

The first research paper was still to come, in 1930 (22). Earlier some other technological discoveries were achieved; for example, VC thermal polymerization (16), phthalic plasticizers (17,18) and polymerization of VC in solution (19). After that it was essential for some basic research to be done. Flumiani (20) determined PVC molecular weight and composition of elements; then Staudinger published his important works on PVC structure. It was Staudinger´s theoretical work that inspired the new technological process, including the quite modern principle of monomer partial conversion to achieve greater uniformity in molecular weight distribution.

Thermal stability became a matter of interest in 1928 when acid neutralizers were applied as PVC prestabilizers (21). Staudinger also referred to PVC degradation in his pioneering works on degradation. Then it became évident that iron present in or around the polymerizing mixture contributes to further catalytic degradation; therefore, an immediate solution was found and enameled equipment applied.

The problem of stabilization divided opinions as no other problem had done before. It was just prior to World War II, and polymeric materials had been seen as having possible strategic importance, especially for Germany, which was known for its inadequate resources of some important raw materials. The Germans, who had scientists who recognized the importance of thermal degradation, did not at the time believe that there was a need for stabilization; therefore, they tended to resolve the problem on a purely technological basis, such as time-temperature conditions of processing and separation of low molecular fractions from the highly polymerized products. At the same time, in the USA and Britain, all the presently-used groups of stabilizers were discovered. Before World War II was over, the world was prepared for all the technological means of PVC production, processing and stabilization.

We can say that this entire briefly-characterized period was the first stage in the development of the field under discussion. If we now try to put all the data into a summarizing sequence, it would appear as follows :

- processing equipment construction,
- monomer synthesis,
- polymerization of monomer,
- various techniques of polymerization technology,
- plasticizers and plastification technology,
- some basic research works,
- stabilizers,
- industrial production.

It seems quite obvious that the structure of the development of knowledge proves the success of practice over theory. It would not be premature if one were to conclude that the first stage, even if very important from a practical point of view, has complicated further development of the field since scientists after World War II have been forced to lay their foundations under already-existing practical applications.

REFERENCES

1. M. Kaufman in **The History of PVC**, London, McLaren, 1969.
2. E.G. Fischer in **Extrusion of Plastics**, London, 1958.
3. H.V. Regnault, **Ann. Chim. (Phys.)**, 58(1835)307.
4. **US Patent** (1836), E.M. Chaffee.
5. H. Kolbe in **Ausfuerliches Lehrbuch der organischen Chemie**, Branuschweig, 1854.
6. T. Hancock in **Personal Narrative of the Origin and Progress of the Caoutchouc or India Rubber Manufacture in England**, London, 1857.
7. **Brit. Patent** 1313/1865, A. Parkes.
8. **US Patent** 133,229 (1872), I.S. Hyatt.
9. E. Bauman, **Liebigs Ann.**, 163(1872)308.
10. F. Wyatt, **Eng. Minning J.**, 58(1894)558.
11. **German Patent** 250,690 (1909), F. Hofman and K. Delbruck.
12. **US Patent** 1,149,577 (1910), K. Gottlob.
13. **German Patent** 278,249 (1912), F. Klatte.

14. **German Patent** 281,887 (1913), F. Klatte.

15. **German Patent** 281,687 (1913), F. Klatte.

16. **Brit. Patent** 156,117 (1918), Traun's Forschungs Laboratorium.

17. **US Patent** 1,398,939 (1920), Eastman Kodak Co.

18. **US Patent** 1,449,156 (1922), US Industrial Alcohol Co.

19. J. Plotnikow, **Z. wiss. Phot.**, 21(1922)133.

20. G. Flumiani, **Z. Elec. angew. phys. Chemie**, 32(1926)221.

21. **Brit. Patent** 336,237 (1928), I.G. Farbenindustrie.

22. H. Staudinger, M. Brunner and W. Feistt, **Helv. chim. Acta**, 13(1930)805.

23. **Brit. Patent** 385,004 (1930), IG Farbenindustrie.

24. **US Patent** 2,068,424 (1931).

25. **Brit. Patent** 401,200 (1932), I.G. Farbenindustrie.

26. K. Krekeler and G. Wick in **Kunststoff-Handbuch**, Munich, 1963.

27. **Brit. Patent** 427,494 (1933), ICI.

28. **Brit. Patent** 451,723 (1934), Carbide and Carbon Chemicals Co.

29. **US Patent** 2,141,126 (1934), A.K. Doolittle.

30. **US Patent** 2,075,534 (1934), F. Groff and M.C. Rees.

31. **German Patent** 679,896 (1935), Deutsche Celluloid Fabrik.

32. **German Patent** 750,428 (1935), A. Wacker.

33. **German Patent** 900,019 (1937), I.G. Farbenindustrie.

34. **US Patent** 2,168,808 (1937), BF Goodrich.

35. **Brit. Patent** 527,408 (1938), B.T.H.

36. **US Patent** 2,227,154 (1938), J.J. Russell.

37. H. Fikentscher, **Angew. Chem.**, 51(1938)433.

38. **German Patent** 61,773 Nc/396.

39. G.M. Kline, W.A. Crouse and B.M. Axilrod, **Modern Plastics**, 17(1940)49.

40. **US Patent** 2,307,092 (1940), V. Yngve.

41. **US Patent** 2,307,157 (1942), V. Yngve.

42. J.M. DeBell, W.C. Goggin and W.E. Gloor in **German Plastic Practice**, Springfield, 1946.

43. **B.I.O.S.** Final Report, 999.

CHAPTER 1

THE CHEMISTRY AND THE STRUCTURE OF THE POLYMER CHAIN

1.1. The chemistry of bonds in PVC

1.1.1. The basic bonds

Poly(vinyl chloride) contains three basic bonds: C-C, C-H and C-Cl. Table 1.1 shows the average bonds´ properties.

TABLE 1.1.

Average properties of basic PVC bonds.

Bond	Bond length, nm	Covalent radii, nm	Bond dipole moments
C-C	0.154	0.077	0
C-H	0.109	0.003	0.40
C-Cl	0.177	0.099	1.46

The C-C bond length may vary from the value quoted above; for example, the values of 0.134 nm and 0.120 nm have been quoted for the double bond in alkenes and the triple bond in acetylenic compounds, respectively. Also, the presence of double bonds in the close neighborhood of the single bond between two carbon atoms will affect the length. Formally the single bond existing in 1,3-butadiene has a length of 0.148 nm.

Similar bond length variations are noted in the case of the C-Cl bond, which may vary in the range of 0.164-0.178 nm. The decrease in bond length has been interpreted in terms of about 10-20% double bond character for the C-Cl bond, owing to the conjugation of an unshared pair of electrons of the chlorine atom with the double bond or aromatic nucleus.

Also, the s-character of the carbon atom results in a bond length change with the highest value for $C(sp^3)$-Cl and the lowest for C(sp)-Cl. Similar changes also affect C-C and C-H bond length, but they are more pronounced in the case of the C-Cl bond than the C-C bond and the latter still more than in the case of the C-H bond. The electronegativity of atoms bonded to the carbon atom is responsible for the further changes in the bond length of C-Cl, but the differences are not large. The same observations concerning the influence of the electronegativity of atoms bonded to the α-carbon atom could be made about the C-H bond, and one should remember that even though the influence on bond length seems insubstantial, the effect can influence ground-state stabilization and therefore modify the rates of reaction and equilibrium constants.

Covalent radii, by their nature, are expected to be as sensitive as bond length to hybridization differences at the carbon atoms, electron-delocalization, electronegativity differences, ionic character and steric effect.

Based on the last values quoted in Table 1.1, we can observe substantial differences in the character of the three bonds discussed. While the C-C bond is non-polar and the C-H bond only slightly polar, the C-Cl bond is very close to an ionic bond in character, which might implicate the reactivity of these bonds and the mechanisms of reactions occurring when they are energetically unstable. Table 1.2 shows the bonds´ dissociation energies.

The same type of bond is broken in each case, with bond dissociation energy varying considerably. For C-H bonds the energy gap equals 79 kJ/mol, and for C-Cl bonds, 66 kJ/mol. In both cases the electron-withdrawing groups contribute to bond energy decrease as it occurs according to the order of the carbon atom. Considerably low dissociation energy of the C-C bond (347 kJ/mol) would suggest that it could be vulnerable to chemical reaction, which would be true if that factor alone could contribute to the reaction rate. Since carbon has a relatively small nucleus with a maximum covalency of four and is able to expand its octet by the use of d-orbitals only with the expenditure of much energy, any transition state involving a fifth bond to the carbon atom has to accommodate the five groups around carbon in excessive electron-density. The geometry of the structural arrangement is thought to be as shown:

$$Y\cdots C\cdots X$$

A consequence of the crowded structure of the carbon atom in the transition state is that bulky groups introduced at the reaction center are generally associated with a decreased reaction rate.

TABLE 1.2.

Bond dissociation energy.

Bond	H, kJ/mol
CH_3-H	435
$(CH_3)_2CH-H$	395
$CH_2=CHCH_2-H$	356
$C_6H_5-CH_2-H$	356
CH_3-Cl	351
$(CH_3)_3C-Cl$	331
$C_6H_5CH_2-Cl$	285

Comparing the C-H and C-Cl bonds, we may expect that the C-Cl bond should dissociate more easily for two reasons: first, because the energy barrier is lower, and second, as the electric charge dislocation facilitates the chemical reaction.

To better understand the subject under discussion and the complexities of the bonds´ nature, one should consider the details of mechanisms of reactions which might occur. At present, a few main mechanisms are considered when elimination reaction occurs.

Elimination of the El type proceeds through the intermediary of the carbonium ion. The sequence of reactions which might occur in such a case could be written for the PVC monomer in the following form:

$$B + -CH_2-\underset{\underset{Cl}{|}}{CH}- \longrightarrow B + -CH_2-\overset{+}{CH}- + Cl^- \longrightarrow BH^+ + -CH=CH- + Cl^-$$

Several factors are known to favor this reaction mechanism. One is the presence of polar solvents in the reacting mixture. Clearly, reactions in solution might follow different routes without polar solvents, as there is a substantial change in energy during reaction and as solvation energy is included. Sometimes these changes in energy are more important for the reaction course than is the strength of forming and breaking bonds. If we also include collision mechanics in the solution, the probability of such collisions and the number of encounters, we can understand that the presence of solvent might play the most important role for the reaction mechanism and for reaction rates (1). On the other hand, the solvent might also contribute to the energy transfer from the molecular fragment in vibrational mode which otherwise would lead to a transitional state, and therefore, it can act in collisional deactivations.

The elimination reaction of the E1 type will also be favored by the good leaving group. Chlorine might be considered as such a group. Finally, the presence of any group at the   position able to stabilize the carbonium ion would contribute to the E1 mechanism's prevalence. This reaction mechanism is especially characteristic for the tertiary and secondary alkyl halides.

The elimination reaction of the E1cB mechanism includes carbanion as the intermediate:

$$B + \text{-CH}_2\text{-CH-} \longrightarrow BH^+ + \text{-}\overset{-}{C}H\text{-CH-} \longrightarrow BH^+ + \text{-CH=CH-} + Cl^-$$
$$\hspace{1.3cm} | \hspace{4.8cm} |$$
$$\hspace{1.3cm} Cl \hspace{4.5cm} Cl$$

The presence of an acidic β-proton is thought to increase the rate of carbanion formation.

A poor leaving group should slow down E2 elimination and favor carbanion formation. Among halogens, fluorine forms the poorest leaving group, and chlorine is second in sequence. The E2 mechanism depends on an energy imbalance introduced in the transitional states described below:

$$
\begin{array}{ccc}
\text{B} & \text{B} & \text{B} \\
\text{H} \quad \delta^- & \text{H} & \text{H} \quad \delta^+ \\
-\text{CH}-\text{CH}- & -\text{CH}-\text{CH}- & -\text{CH}-\text{CH}- \\
\qquad \text{Cl} & \qquad \text{Cl} & \qquad \text{Cl}
\end{array}
$$

Carbanion character   Central character   Carbonium ion character

The character of the transitional state will determine the degree of C-H and C-Cl bond breaking. In the course of the elimination reaction, any group formed or any change in reaction conditions which might affect the charge distribution or activation process favors the particular direction of the elimination process. In the case of the PVC monomeric fragment, the leaving group, that is, HCl, will stabilize the carbanion structure in relation to the carbonium ion and thus increase C-H bond breaking. For the carbanion transition state the β-substituants of the electron-withdrawing character (e.g, double bonds) should both increase the carbanion character and favor C-H bond breaking. Reacting base B might also contribute to the reaction mechanism due to its nucleophilicity.

The E2C transition states have the following structure:

$$
\begin{array}{c}
\text{H} \cdots \text{B} \\
-\text{HC}-\text{CH}- \\
\text{Cl}
\end{array}
$$

E2C elimination is favored by dipolar aprotic solvents such as DMF and any good leaving group (Cl). Also, substrates having tertiary carbon atoms will undergo changes according to this mechanism more easily.

Another mechanism, known in the literature under the name of four-center HX elimination, might help to explain the course of the PVC degradation process. This mechanism was proved on the basis of studies of unimolecular elimination of HX from alkyl halides. The possibility of partial charge separation was proved quite early (2). Further studies of the reaction and its activation energy parameter for many alkyl monohalides confirmed the polar nature of reaction and the existence of surface catalysis (3). The essence of the above mechanism can be explained by the following equation:

$$\begin{array}{c} R_1 \\ \diagdown \\ {}^{R_2} \diagup \end{array} C - C \begin{array}{c} R_3 \\ \diagup \\ \diagdown \\ {}^{R_4} \end{array} \rightleftarrows \left[\begin{array}{c} R_1 \\ \diagdown \\ {}^{R_2} \diagup \\ {}_{\delta^-} \end{array} \overset{\delta^+}{C} \overset{\delta^-}{\cdots} \overset{}{C} \begin{array}{c} R_3 \\ \diagup \\ \diagdown \\ {}^{R_4} \end{array} \right] \rightleftarrows \begin{array}{c} R_1 \\ \diagdown \\ {}^{R_2} \diagup \end{array} C = C \begin{array}{c} R_3 \\ \diagup \\ \diagdown \\ {}^{R_4} \end{array} + HX$$

It has been established that methyl and other n-alkyl groups bound to the α-positioned carbon atom participate in the stabilizing effect. The chlorine atom affects the formation of the positive charge on the carbon atom to which it is bound, but it does not contribute to the formation of a negative charge on the neighboring β-carbon. The effect just described will contribute to C-X bond energy, which, together with H-X bond dissociation energy, determines the rate of initiation. The presence of HX can promote autocatalytic reactions. Overall, one might expect a dehydrochlorination reaction of the first order to take place.

Molecular elimination of HX is usually preferred in the case of primary and secondary carbon atoms, while the radical route might prove to be essential when halides are bound to tertiary carbon atoms. The activation energy in the four-center elimination process is drastically decreased when methyl, vinyl, aromatic groups or oxygen are bound to a carbon atom which forms a C-X bond at the same time. If one of these groups is bound to a carbon atom forming a C-H bond, the effect is negligible or not as strongly pronounced as in the other case. This might suggest that if one considers the regular distribution of the C-X bonds at each second carbon atom, which is generally the case in PVC, one may expect reaction to proceed along the chain until the first irregularity appears, which might be why reaction is terminated for that center.

Having completed this brief listing of the major mechanisms believed to be essential for explaining the nature of the HX elimination process, we still need to describe the reaction which involves the substitution of chlorine by other groups in the course of the same reaction. Two well-known mechanisms of substitution have been used in organic chemistry for years: the S_N1 and S_N2 types. For our studies it is essential to notice the differences between both mechanisms in respect to reactions which may occur if segments of the PVC chain are thermodynamically unstable and therefore able to react. As we concluded above, the carbon atom, because of its four-valent character and its need to form a crowded transition state, including accommodation of five groups, is not the easiest one to undergo substitution reaction. Referring that to the mechanism of S_N2 type, which includes such a transition state, one might expect that the substitution of chlorine by any other group according to the S_N2 mechanism would be more likely to occur in the case of secondary than tertiary carbon atoms in α- and β-positions to the reaction center. Thus, all the branches available at the reaction site will form unfavorable geometric hindering effects for the reaction to be performed. This is in complete contradiction to the conditions which might favor reaction of the S_N1 type, which is more likely to occur when a C-Cl bond is formed with a tertiary carbon atom. On the other hand, electron-attracting groups (e.g., double bonds) will promote the S_N2 mechanism and retard S_N1. Therefore, if substitution of unstable chlorine bound to a tertiary carbon atom takes place, it is likely done through a mechanism of the S_N1 type at the beginning of the degradation process before unsaturated fragments are formed.

The S_N1 mechanism includes the formation of a carbonium ion in the first step of the reaction. Therefore, it is favored by a good-leaving group, polar solvents and α-substituants, which stabilize the carbonium ion. The stabilizing α-substituant prolongs the accessibility of ions for reaction. One peculiarity in the S_N1 mechanism is potentially useful for explaining the PVC degradation mechanism. Catalysis of PVC dehydrochlorination by metal salts normally appears to be S_N1-like, with a transition state as follows:

$$\left[\begin{array}{c} -\overset{.}{C}H \!-\! CH_2 - \\ \vdots \\ \overset{.}{C}l \\ \vdots \\ \overset{.}{M} \end{array} \right]^{+}$$

The possibility of the catalytic influence of metal salts allows for substitution of chlorine in a more advanced stage, e.g., when a C-Cl bond under consideration is already in the neighborhood of a double bond or bonds formed in former acts. The last remark on the catalytic effect of metal salts and the earlier one on the catalytic effect of HCl bring us closer to the final topic in this section: the possibility of inter- and intra-molecular hydrogen bonding and its possible eventual implications (4).

Formation of the following hydrogen bonds is theoretically possible:

(a)
```
        -CH——CH-
         |    |
         Cl   H
              .
              .
             ClMCl
```

(b)
```
        -CH——CH2-
         |
         Cl
         .
         .
         H
         |
         Cl
```

(c)
```
        -CH——CH-
         |    |
         Cl· · ·H
```

(d)
```
         Cl    H
         |     |
        -C——— C-
         |     |
         H     H
         .
         .
         Cl
         |
        -CH——CH2-
```

One should take into consideration that in the most cases chlorine is the most active halogen in hydrogen bond formation. Usually the energies of hydrogen bonds with chlorine are in the range of 12 to 42 kJ/mol, which is designated as normal, and some of these bonds have energy above 42 kJ/mol, which is a strong hydrogen bond.

Taking into consideration structure (d), which suggests the formation of an intermolecular hydrogen bond between two chains, the C-Cl bond has a pronounced electron-withdrawing effect upon other bonds associated with the same carbon atom; therefore, the hydrogen atom may become sufficiently acidic so as to act as an electron-acceptor atom in a hydrogen bond. Considering the structure (c), it must be kept in mind that the most stable forms of hydrogen bond occur when the proton-donor and proton-acceptor atoms are colinear. "Bent" hydrogen bonds can occur, but they are weak. Also, the steric effect arising from constraints such as double bonds within the molecule might prevent hydrogen from approaching a chlorine atom. Conversely, the hydrogen group linked to the sequence of conjugated double bonds is a much stronger proton donor.

Reconsidering the mechanism of the four-center HX elimination process from the point of view of the hydrogen bonds available, we see that an intra-molecular hydrogen bond can secure the negative charge on the β-carbon atom and therefore promote the reaction to follow that mechanism.

In respect of structures (a) and (b), one might say that similar bonds have been observed in some compounds, and as catalytic effects are observed in heterophasial systems which are, respectively, liquid-solid and liquid-gas, it is possible that they might occur due to the former hydrogen bond formation which weakens the initial bonds and permits the split-off of hydrogen or chlorine, respectively.

This brief survey of reaction mechanisms is quoted for the further discussion of PVC decomposition. The principles of reactions have been formed on the basis of simple compounds, because for PVC such data does not yet exist. The scope of PVC decomposition concerns small molecules (e.g., HCl, stabilizers, etc.) in many cases. In others, the reactions occur due to a single collision that does not involve the whole chain but only its small fragment, which is in proper energetic condition and might behave like a moiety of low molecular weight, especially if the collision occurs with a small molecule.

1.1.2. The bonds available in the chain from
contaminations and formed in the process of thermal degradation

Apart from regular units in the PVC chain, some other bonds are
formed in polymerization, during storage, during processing and
in the use of polymer. An unrestricted variety of such bonds
could be listed, but the main ones comprise C-S, C-O, C=C and C=O
bonds. All are formed during processing and use, and the last
three can be found directly after polymerization is completed.
To facilitate our studies on polymer behavior, we should find the
mechanism of these bonds´ formation and how they modify the
polymer, i.e., what is to be expected if polymer contains these
bonds. The answer to the first question is not of primary
concern here, for we shall investigate it throughout this book,
as it is the main problem so far as the thermal degradation of
PVC is concerned.

C=C bond

Due to creation of a double bond, the distance between both
atoms involved is decreased considerably from an average value of
0.154 nm to 0.133 nm. The bond energy is also changed as it
varies from 250 to 603 kJ/mol for a C-C bond, and it is equal to
720 kJ/mol for ethylene. The double bond is rigid and has less
freedom in rotation. If, for instance it were rotated by 90°,
the 2p orbitals would be perpendicular to each other and no
molecular orbital would be formed. It is therefore evident that
only rotation by 180° is not forbidden. When another double bond
is formed in conjugation, as for example, in a butadiene
molecule, both lengths and energies of bonds are further
changed. Neither single C-C bonds nor double bonds are
comparable to either ethane or ethylene; moreover; the total
energy in butadiene is not equal to the sum of the energies of
one single and two double bonds. This problem will be discussed
in more detail in the following chapters. These observations are
intended to point out that C=C bond strength is quite high, the
molecule is flat in the double bond area, and the change in
spatial arrangement is difficult and restricted to two positions
only adequate to rotation by 180°.

One interesting feature concerning the double bond is its rearrangement. There are several possibilities for such a process to proceed. One of them, allylic shift, occurs due to the nucleophilic substitution at an allylic carbon. When allylic substrates are subjected to nucleophilic action due to the S_N1 mechanism, two products are usually obtained:

$$R-CH=CH-CH_2X \xrightarrow{Y^-} R-CH=CH-CH_2Y + R-CH(Y)-CH=CH_2$$

Similar substitution with rearrangement may also take place by an S_N2 mechanism if α-substitution sterically retards the normal S_N1 mechanism. Electrophilic substitution can also be accompanied by double bond shifts, in which case it is analogous to the S_E1 mechanism:

$$-\overset{|}{C}=C-\overset{|}{\underset{X}{C}}- \longrightarrow [-C=C-\overset{|}{\underset{-}{C}}- \longleftrightarrow \overset{-}{-C}-C=C-] \xrightarrow{Y^+} -\overset{|}{\underset{Y}{C}}-C=C-$$

For nucleophilic substitution to proceed, strong bases (such as $NaNH_2$) are needed, which have a lower probability to be present in a PVC environment, unlike proton and Lewis acids, which are needed for substitution and rearrangement of the double-bond by the S_E1 mechanism. Still another possibility for rearrangement may explain changes at chain ends. Sigmatropic hydrogen migration proceeds at elevated temperatures or in the presence of radiation, according to the following equation:

This rearrangement is configuration specific, and only the molecules able to resume cisoid conformation can be subjected to such change. 1,4 conjugated dienes can be rearranged by undergoing the Diels-Adler reaction:

Many conditions must be fulfilled for such a reaction to occur. Diene has to be in cisoidal conformation while effective dienophile (Z-CH=CH-Z´) should have at least one of the following groups (Z or Z´): CHO, COR, COOH, COOR, COCl, COAr, CN, NO_2, Ar, CH_2OH, CH_2Cl, CH_2NH_2, CH_2CN, CH_2COOH, Cl, C=C. Finally, the lowest unoccupied molecular orbital of one reactant should overlap with the highest occupied molecular orbital of the other. It is interesting to note that cyclodienes or linear trienes can be interconverted by treatment with either UV light or heat:

This reaction, called electrocyclic rearrangement, is the most probable reason for benzene formation in the degradation process.

Additions to double bonds may proceed with numerous reagents, but only some reactions are relevant for the field under discussion. In conjugated systems electrophilic addition produces a mixture of two possible structures:

It is important to notice that if a 1,4 addition product is formed, the conjugation is definitely disrupted, which is not necessarily the case with a 1,2 addition product. Furthermore, a double bond in conjugation with a carbon-hetero multiple bond (e.g, C=O) lowers the addition rate and provides competition from the 1,4 addition. Addition of carboxylic acids to a double bond is a process of electrophilic mechanism catalyzed by proton or Lewis acids (e.g., $ZnCl_2$). The addition of thiols to double bonds is by electrophilic, nucleophilic or a free-radical mechanism, and proceeds faster when an acid catalyzer is present.

The other important group of reactions, so far as the C=O double bond is concerned, includes oxidation. First of all, oxidation usually takes part in the transfer of the hydrogen atom:

$$RH + Cl^{\cdot} \longrightarrow R^{\cdot} + HCl$$

which should explain the higher dehydrochlorination rate of PVC in the presence of oxygen. The presence of a radical allows for further changes:

$$R^{\cdot} + O_2 \longrightarrow R\text{-}OO^{\cdot}$$

In the presence of atmospheric oxygen, there is the possibility of an autooxidation reaction catalyzed by light in which hydroperoxides are produced:

$$RH + O_2 \longrightarrow ROOH$$

and a C-O-OH group formed. Photooxidation of dienes may produce internal peroxides:

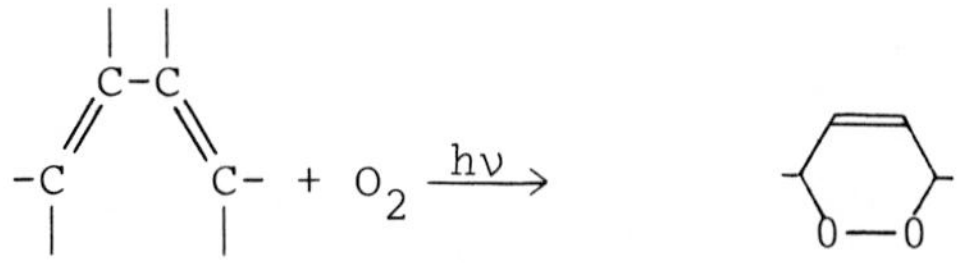

while singlet oxygen can also react with double bonds in another way to give a dioxetane intermediate, which usually cleaves to aldehydes or ketones:

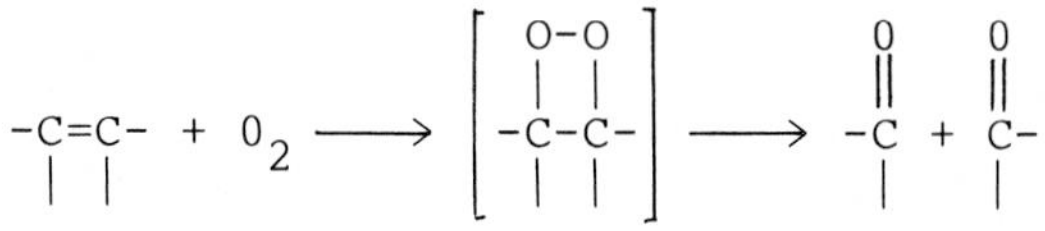

Double bonds react with ozone:

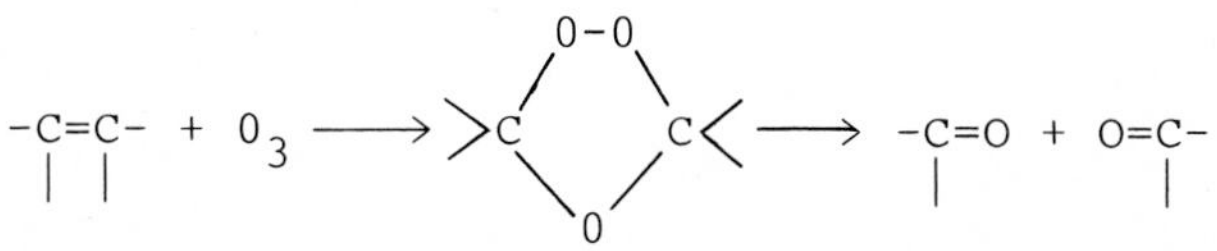

to form aldehydes or ketones. The last step is well catalyzed by the presence of zinc.

From this brief review we can see that there are a wide variety of possibilities in which double bonds can be affected, and the products formed are usually either photolytic reaction sensitizers or products which increase PVC instability during either thermal or radiation treatment.

C=O bond

The carbonyl group is considered to be one of the most reactive, but the types of reactions usually do not apply to our topic. Three features are important for consideration in the PVC degradation processes. One is connected with the electronegativity of oxygen, which contributes to the strong electron-withdrawing property of this bond, which, in consequence, leads to a weakening of bonds adjacent to the carbon atom forming a carbonyl group; therefore, neighboring C-H and C-Cl bonds have a higher probability of reacting. The second, and probably the most essential property of the carbonyl group, is that it can absorb light in the 230 to 330 nm region, which results from the $n \to \pi^*$ singlet-singlet transition. The excited ketone can then cleave:

$$R'\text{-}\underset{\underset{O}{\|}}{C}\text{-}R \xrightarrow{h\nu} R'\text{-}\underset{\underset{O}{\|}}{C}{}^{\bullet} + R^{\bullet}$$

This reaction is called Norrish Type I cleavage. If a ketone has a γ-hydrogen, it can cleave according to Norrish Type II cleavage:

$$R_2CH\text{-}CR_2\text{-}CR_2\text{-}\underset{\underset{O}{\|}}{C}\text{-}R' \xrightarrow{h\nu} R_2C{=}CR_2 + R_2CH\text{-}\underset{\underset{O}{\|}}{C}\text{-}R'$$

One can easily imagine that both reactions are rather harmful, and they can start many secondary processes discussed above. Ketones may also take part in the Paterno-Buechi reaction:

$$\underset{\underset{O}{\|}}{\text{-C-}} + \underset{\substack{\| \\ \text{-C-}}}{\text{-C-}} \xrightarrow{h\nu} \underset{\substack{| \quad | \\ O\text{-C-} \\ |}}{\text{-C-C-}}$$

The resultant oxetanes serve as cross-linkages between two polymer chains, which may help us to understand more about the nature of the crosslinking process.

C-O bond

We should still refer here to the reaction between the double bond and carboxylic acid, as this is possibly one of the pathways by which a C-O bond can be formed in the chain. In the double bond addition reaction, there is a theoretical possibility that carboxylic acid can substitute into a chain by reaction with a double bond that has already been formed. Such an addition reaction is normally acid catalyzed, and the type of catalyzer plays an essential role. If the acid is proton-donating such as, for instance, HCl, the reaction goes according to the scheme shown above, but it has an entirely different pathway if it is catalyzed by Lewis acids, for example, $ZnCl_2$. In this case, an acylonium ion is created which attacks the double bond-forming ketone. In the consequence of the ketone forming as discussed above, further degradation may ensue due to light absorption and radical and crosslink formation. Naturally, the presence of carboxylic acid rest should mainly be attributed to the fact that the reacting mixture contains both the salt of carboxylic acid and the halogen attached to the polymer chain. In this case, it is to be expected that, due to the S_N2 mechanism, carboxylic acid rest should replace chlorine in the chain.

Finally, we are interested in what would happen to a C-O bond in more advanced stages of degradation when an acid catalyzer (HCl) is present in excessive quantities. It may be predicted that the normal hydrolysis process should occur, leaving an hydroxyl group in the chain. The process is related to temperature, so the compound may also undergo elimination:

$$-\underset{\underset{O=\underset{R}{C}}{\overset{H}{C}}}{\overset{|\ \cdot\ |}{C}}-\ \longrightarrow\ \underset{RCOOH}{\overset{|\ \ |}{-C=C-}}$$

Both processes still do not explain some observations in advanced stages of PVC degradation when "rapid blackening" occurs. It seems feasible to suggest that the following changes take place. On one hand, the carboxylic acid can be a leaving group, and therefore an additional double-bond is formed. The acid rest existing in the chain can also be hydrolized the other way, leaving behind an hydroxyl group in the chain. In this case we arrive at a structure containing the double bond in the neighborhood of the hydroxyl group because the former addition was most probable in a conjugated sequence. Then we should add that in advanced stages of degradation, double bond rearrangement can easily occur since both proton and Lewis acids are present in excessive quantities. Finally, due to keto-enol tautomerism:

$$-\overset{\displaystyle|}{\underset{\displaystyle OH}{C}}=\overset{\displaystyle|}{C}- \;\rightleftharpoons\; -\overset{\displaystyle|}{C}-\overset{\displaystyle|}{\underset{\displaystyle O}{C}}-$$

we can easily obtain carbonyl structures in the chain which can still undergo the chemical changes discussed above.

C-S bond

Sulphur is less electronegative than oxygen and even chlorine. If one compares the charge distribution in the C-O and C-S group for methanol:

$$\overset{-0.322}{C}\text{———}\overset{-0.741}{O} \qquad \overset{-0.841}{C}\text{———}\overset{-0.109}{S}$$

it is evident why sulphur incorporated into the chain does not cause any instability. Also, the relatively high valence shell size contributes to a steric effect that protects against further chemical changes. Generally, we can regard sulphides as stable moieties which, if hydrolized or oxidized, form compounds that do not promote further degradation of polymer.

1.2. The irregular segments of the chain

The long-ago observed PVC thermal instability suggested to some authors that it is due to the atypical bonds present in PVC molecules. The above discussion was meant to characterize the chemical properties of such bonds, while the discussion below is intended to find out how numerous they are, and what is the process of their origination. All this is to prepare the material for discussion of PVC degradability under varying conditions which is to be included in Chapters Two and Three.

1.2.1. Unsaturations

Several methods can be used to study the unsaturated structures present in PVC. The main method is based on bromination, which gives the total number of double bonds present in polymer. Ozonolysis, in turn, coupled with molecular weight change determination, gives the number of double bonds which are internally present in the polymer chain. The difference in reading gives double bonds at chain ends. Numerous investigators have studied the chemical structure of chain ends. The most often applied technique is based on the ^{1}H-NMR method.

Petiaud (5) and Schwenk (6) were able to record signals assigned to the following structures:

$$-CH_2-CH=CH-CH_2Cl$$
$$-CHCl-CH=CH-CH_2Cl$$

Caraculacu (7), using Fourier transform ^{1}H-NMR, found a series of five signals, three of which were assigned to the following structures:

$$-CH_2-CHCl-CH=CH_2$$
$$-CH_2-CHCl-CCl=CH_2$$
$$-CH_2-CH=CCl-CH_2Cl$$

Hjertberg (8) confirmed the presence of Schwenk´s structure:

$$-CHCl-CH=CH-CH_2Cl$$

and found that the other groups included at chain ends contain saturated groups: $-CHCl-CH_2Cl$ and $-CH_2-CH_2-Cl$. The presence of all these groups is discussed by Braun (9). Hjertberg (8) showed the probable path of their formation, which is given by the following equations:

$$-CH_2-\overset{\bullet}{C}HCl \xrightarrow{VC} -CH_2-CHCl-CHCl-\overset{\bullet}{C}H_2 \xrightarrow{1,2-Cl\ migration}$$

$$-CH_2-CHCl-\overset{\bullet}{C}H-CH_2Cl \xrightarrow[-Cl]{} -CH_2-CH=CH-CH_2Cl$$

$$Cl^{\bullet} + CH=CHCl \longrightarrow CH_2Cl-\overset{\bullet}{C}HCl \xrightarrow{VC} CH_2Cl-CHCl-$$

Also, according to Hjertberg (8), the isolated allylic structure can easily be converted to a diene structure via an ionic mechanism:

$$-CHCl-CH_2-CH=CH-CH_2Cl \xrightarrow[-Cl]{} [-CHCl-CH_2-CH=CH-\overset{+}{C}H_2 \longleftrightarrow$$

$$-CHCl-CH_2-\overset{+}{C}H-CH=CH_2] \xrightarrow[-H^+]{} -CHCl-CH=CH-CH=CH_2$$

The presence of internal double bonds is explained by copolymerization with an acetylenic impurity present in monomeric vinyl chloride (10).

There is a large disparity between the number of internal double bonds and chain-end double bonds. Internal double bonds are present in amounts of 0.1-0.8/1000 monomer units, which is why they cannot be studied effectively by NMR. The number of internal double bonds depends on monomer pressure (8), but it is not related to the molecular weight of polymer. Chain-end double bonds are definitely more numerous, and the usual range detected by various authors is between 1.2-4.8 double bonds per 1000 monomer units. Hjertberg (8) explained that the number of double bonds at chain ends is closely related to the molecular weight of polymer.

Minsker (11, 12) says that allylic chlorine does not exist in normal polymer as it is immediately oxidized during production, drying and storage to form:

$$-CH_2-CH=CH-CHCl- + O_2 \longrightarrow -C-CH=CH-CHCl- + H_2O$$
$$\underset{O}{\overset{\|}{}}$$

Caraculacu (13), discussing this assumption, asked

why the model substances consisting of 4-chlorohexene-2 together with its transposed derivative 2-chlorohexene-3, are not oxidized to corresponding keto-chloroallylic structures even after one year of standing in air or oxygen, since only a partial dehydrochlorination and the consequent formation of the hexadiene has been observed ?

1.2.2. Branches

Studies of PVC behavior in solution have developed interest in the branching characteristics of the polymer chain (14, 15). Later, Andersson explained that the abnormal behavior of solution originates from aggregates (16) not, as was suspected, long branches, but at the same time interest was developed in investigating branches for the sake of explaining a mechanism of thermal degradation. In the meantime, Rigo (17) proposed the mechanism of chloromethyl branch formation, which was followed by two other mechanisms (18, 19). More recently, Starnes (20) gave what seems to be the final proof that Rigo´s mechanism is correct:

$$-CH_2-CHCl-\overset{\bullet}{C}H-CH_2Cl \xrightarrow{\text{VC}} -CHCl-CH_2-\underset{\underset{CH_2Cl}{|}}{CH}-CHCl-CH_2-$$

Other works suggested that there are also 2-chloroethyl (21) and 2,4-dichlorobutyl branches (18) to be taken into account which have the following formula:

$$\underset{-CH_2-CCl-CH_2-}{\overset{CH_2-CH_2Cl}{\overset{|}{}}}$$

$$\underset{-CH_2-CCl-CH_2-}{\overset{CH_2-CHCl-CH_2-CH_2Cl}{\overset{|}{}}}$$

In the meantime, Starnes (19) developed an excellent technique of branch point determination by reductive-dehalogenation of PVC using either Bu_3SnH or Bu_3SnD for ^{13}C NMR spectroscopy. The route of 2-chloroethyl branch formation was confirmed (19):

$$-CH_2-CHCl-CH_2-\overset{\bullet}{C}HCl \longrightarrow -CH_2-\overset{\bullet}{C}Cl-CH_2-CH_2Cl \xrightarrow{VC} -CH_2-\underset{\underset{CH_2-CH_2Cl}{|}}{C}Cl-CH_2-CH_2Cl$$

which is obtained after intramolecular hydrogen-transfer, by the addition of monomer. A detailed study of butyl branches was also done by Starnes (22) using the same technique. The mechanism is not yet known completely, but there is a suggestion that butyl branches are formed as a result of a radical "backbiting" reaction.

Bowmer (23) recently characterized branches by γ-irradiation of reduced polymer, i.e., the technique used for polyethylene was applied to PVC after it was converted to a similar chemical structure. All three branches were detected in an irradiated polymer sample closed in an ampoule when volatile products were analyzed by gas chromatography. No matter the polymer type, 2-chloromethyl branches always account for more than 50% of the branchings.

The branches are in the following range:

2,4-dichlorobutyl	0.6-1/1000 mers (22)
2-chloroethyl	0.4-2.4/1000 mers (13)
chloromethyl	4-6/1000 mers (13)

Finally, we should see the difference between the chloromethyl branch and two others, since the tertiary carbon at the chloromethyl branching point is bound to hydrogen, while in two other cases it bonds with chlorine (mainly).

1.2.3. Oxygen-containing groups

During vinyl chloride polymerization in the presence of oxygen, groups containing oxygen are formed in polymer (24, 25). The peroxides initially formed during polymerization are not stable, and they yield HCl, CO and formaldehyde. Braun (9), who analyzed the aqueous phase during the polymerization, definitely confirmed that the amount of HCl formed depends on oxygen concentration; also, the number of carbonyl groups in polymer depends on the amount of oxygen in the polymerization environment, as shown on Fig. 1.1.

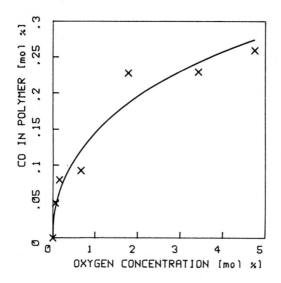

Fig.1.1. Carbonyl group formation during polymerization in the presence of oxygen. (Data from Ref. 9.)

The molecular weight of polymer is also affected by the presence of oxygen and the internal double-bond concentration (Fig. 1.2.). What is interesting in comparing both graphs is that the number of double bonds is several times lower than that of the carbonyl groups. This observation cannot, therefore, exclude the possibility that Minsker is right in respect to the presence of keto-allyl groups, but at the same time, it also does not support his theory.

28

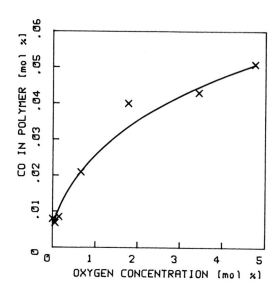

Fig.1.2. The formation of internal double bonds during VC polymerization in the presence of oxygen. (Data from Ref. 9.)

1.2.4. The head-to-head structures

Here is another example of our lack of information on the essential parameters of PVC structure that may affect its quality. Only Mitani (26) suggested that six to seven head-to-head structures per 1000 mers can be found in PVC, and more recently, Crawley (27) synthesized head-to-head PVC from chlorinated cis-1,4-polybutadiene and determined its thermal properties, which will be discussed in Chapter 2. Otherwise, we can find in the literature only assumptions that the presence of this structure might be essential for PVC thermal stability.

1.2.5. Initiator and transfer agent rests

This problem was never the subject of detailed investigation, although it is suggested from time to time that if such groups are present at chain ends, they may affect polymer thermal stability (13, 19). Some research work has been done on the presence of transfer agent rests, but they were not found in the polymer chain (13). It seems very important to analyze these possibly labile structures in the polymer chain in order to know if they should be taken into consideration when thermal stability is concerned.

1.2.6. Total labile chlorine, tertiary carbon chlorine and allylic chlorine

The above-listed names were created in consequence of the analytic procedure used in the determination of labile structures in the polymer chain. Adequate methods will be discussed in Chapter 6, and here we will only decode the names frequently used in research papers. Total labile chlorine refers to the number of chlorine atoms able to undergo phenolysis, and it should include the sum of allylic chlorine and tertiary carbon-bound chlorine. When phenolysis is done after bromination, we should be able to measure the concentration of chlorine bound to the tertiary carbon atom, which should also agree with the number of long branches determined by NMR. Usually the number of long branches is higher than the number of tertiary carbon chlorines as determined by phenolysis after bromination, which suggests that the chlorine is not always bound to carbon at the long branch. The difference in the phenolysis reading before and after bromination gives us the concentration of allylic chlorine. Finally, the difference between the number of double bonds as determined by bromination and chain scissions after ozonolysis gives the number of double bonds at chain ends.

1.3. Structure of the chain

Few authors have attempted to comment on this subject, especially because of the difficulties in either measuring techniques or quantum mechanical calculations. PVC, being a polymer of a low degree of crystallinity, cannot be studied effectively by the usual methods applied to crystalline materials. Recent research, however, has given some interesting results, which will be commented on below.

1.3.1. Molecular weight of polymer (chain length)

Ideally, we should like to see here a straightforward correlation between either the molecular weight of polymer or the distribution of molecular weight and thermal stability of PVC. In early studies such attempts were made (28), resulting in rather more confusion than clear explanation of the effect of molecular weight on PVC thermal stability. Table 1.3 shows some of the data quoted from these studies.

TABLE 1.3.

The number-average molecular weight versus
PVC dehydrochlorination rate. (Modified from Ref. 28.)

M_n	Polymerization temp., K	Dehydrochlorination rate, $10^6 dx/dt$, s^{-1}
55,000	298	1.66
95,000	273	1.78
142,000	253	1.56
136,000	233	1.61
93,000	213	2.33
79,000	195	1.87

The same reference (28) contains data on the thermal stability of PVC obtained by fractional precipitation with a tetrahydrofuran-water system. The results are plotted in Fig. 1.3.

The first set of data presented by Table 1.3 does not indicate any correlation between PVC molecular weight and its thermal stability, while the data in Fig. 1.3 suggest that there is a range of molecular weight most appropriate to achieve maximal thermal stability. In discussing the results obtained, the authors (28) doubted the possibility of a zipper dehydrochlorination mechanism or the effect of tertiary chlorine and stereoregularity on the PVC degradation rate. Recently, a similar set of data was presented in a more current publication

(29) (Table 1.4).

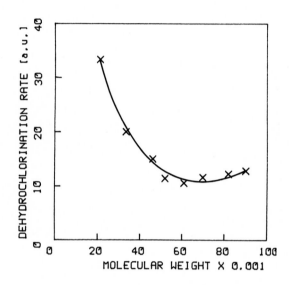

Fig.1.3. The relationship between the molecular weight of PVC (M_n) and its thermal dehydrochlorination rate. (Data from Ref. 28.)

TABLE 1.4.

The PVC degradation rate versus polymer molecular weight.
(Modified from Ref. 29.)

Intrinsic viscosity,ml/g	A_{1428}/A_{1434}	Degradation rate x 10^3/min
60.3	1.04	4.34
95.4	1.13	4.83
168.0	1.19	3.91
193.0	1.27	7.67
138.6	1.38	33.50
144.5	1.05	3.65
402.0	1.27	2.98
114.5	1.53	11.04

Regardless of the method of sample preparation, we can again see that molecular weight alone does not explain polymer instability, as the dehydrochlorination rate seems to correlate better with the infrared absorbance ratio, being a relative measure of the tacticity content, than it does with the molecular weight of polymer. Hjertberg's and Sörvik's work (8) supplies other important data regarding the effect of molecular weight on PVC thermal stability. Fig. 1.4. shows data comparing the molecular weight of polymer with the number of double bonds for samples prepared in the laboratory and suspension polymer samples extracted with various solvents.

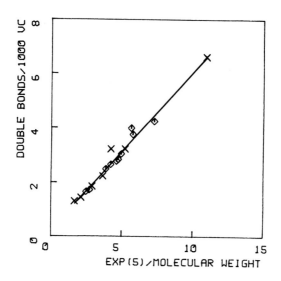

Fig.1.4 The number of double bonds versus molecular weight of polymer. (Modified from Ref. 8.)

Fig. 1.5. Shows similar data for labile chlorine as determined by phenolysis (30). From both graphs it is evident why it is not possible to find a simple correlation between molecular weight of PVC and its dehydrochlorination rate. If such a possibility had existed, it would mean that molecular weight, labile chlorine, tacticity and double bonds have an equivalent effect on PVC degradation rate, which cannot be the case.

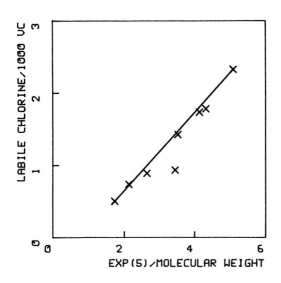

Fig.1.5. The amount of labile chlorine versus molecular weight of polymer. (Modified from Ref. 8.)

The problem is well characterized by Hamielec studies (31) and set of model equations by Graessley (32). In the case of molecular weight controlled mainly by transfer reactions (small amount of polymer produced by termination), the molecular weight and branching development can be calculated as follows:

$$dQ_0/dX = C_M$$

$$dQ_1/dx = 1$$

$$dQ_2/dx = 2\frac{1 + C_p Q_2}{C_M + C_p X/(1 - X)}$$

$$\frac{d(Q_0 \bar{B}_N)}{dx} = C_p\{X/(1 - X)$$

$$C_M = k_{fm}/k_p = C_{MO}\exp\{B_M(1/V_F - 1/V_{FcrM})\}$$

$$C_p = k_{fp}/k_p = C_{pO}\exp\{B_p(1/V_F - 1/V_{Fcrp})\}$$

$$\bar{M}_N = M_m Q_1/Q_0 = 62.5X/Q_0$$

$$\bar{M}_W = M_m Q_2/Q-(1) = 62.5Q_2/X$$

$$\bar{\lambda}_N = 1000M_R(\bar{B}_N/\bar{M}_N)$$

where:

X - monomer conversion,

$\bar{B}_n$ - number-average of branch point per polymer molecule,

k_{fm} - transfer-to-monomer rate constant,

k_p - propagation rate constant,

k_{fp} - transfer rate constant,

$\bar{M}_N$ - number-average molecular weight of polymer,

$\bar{M}_W$ - weight average molecular weight of polymer,

V_F - transfer rate,

$\bar{\lambda}_N$ - number of long branches per 1000 monomer units,

M_R - molecular weight of the repeat unit,

B_M, B_p, V_{FcrM}, V_{Fcrp} - parameters which can be found by fitting the rate and the molecular weight data.

Form the above model and data derived from calculations (31), one can see that similar to experimental results quoted above (fig. 1.5), the number of long branches is inversely proportional to the polymer molecular weight. The molecular weight of polymer applied in recent processing methods expressed by $\bar{M}_n$ varies in the the range of 35,000 to 80,000, and during the last few years, due to the improvements in the polymerization process, the lower limit was decreased from 45,000 to 35,000, which was no possible for a long time. The distribution of the molecular weight of PVC can be seen from Fig. 1.6 (33) which shows how a wide range of molecular weight should be expected which causes another difficulty in the estimation of the effect of molecular weight of polymer on its thermal stability.

Summarizing the above discussion, one should say that, at the moment, there are difficulties in concluding the effect of molecular weight on PVC thermal stability mainly due to the unresolved problem of the effect of various irregularities on PVC dehydrochlorination kinetics. It is obvious that by decreasing the molecular weight, we should promote degradation reactions due to the increased reaction probability as the influence of the physical state barrier lowers in parallel to the chain length. On the other hand, the lower molecular weight of polymer allows less severe conditions of processing and helps to decrease heat generated on shearing.

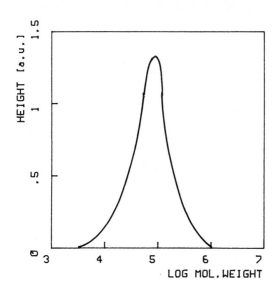

Fig.1.6. PVC molecular weight distribution. (Modified from Ref. 33).

1.3.2. Configuration and conformation of the polymer chain

The first-order structure of the polymer chain is determined by mers arrangement, configuration and conformation. The mers arrangement includes head-to-head and head-to-tail structures that have already been discussed. By configuration one should understand the spatial arrangement of atoms, and groups formed by them, in reference to a particular point of a chain-like center of asymmetry or chain backbone. In the first case, different configuration causes the optical isomerism for which there has to be asymmetric carbon available in the side group (this does not apply to PVC). In the second case, we are dealing with a particular type of stereoizomery called tacticity, which is common so far as PVC is concerned. Polymer tacticity exhibits three basic structures:

1. isotactic, in which all the groups are on the same side of a chain (in PVC, all the -Cl groups);
2. atactic, in which these groups are randomly and irregularly distributed;
3. syndiotactic, in which groups, e.g., Cl, are regularly

distributed on both sides of the chain or its segment in alternative arrangement.

Comparing optical isomerism with tacticity, one should mention that, while physical properties of optical isomers are the same (except light polarization), polymer tacticity affects physical and mechanical properties of polymer such as strength, elasticity, light absorption, phase transition temperature, crystallinity and so on.

Isotactic and syndiotactic segments can be expressed by the following structures:

Configurational content can be studied by high resolution NMR, IR and Raman spectroscopies. NMR spectroscopy was used by Sörvik (34) who distiguished six bands assigned to syndio, isotactic and combined sequences. The chief drawback of this method is that it can be used only for polymers in solution, while highly syndiotactic polymers are insoluble, and moreover, they may change their initial properties in solution and under conditions of dissolution.

IR spectroscopy is the most frequently applied method for tacticity determination. In this method, by comparing the intensities of two bands, one can obtain an index of tacticity. Most frequently used ratios are A_{615}/A_{690} (35) and A_{1428}/A_{1434} (36). The sample preparation method plays an important role here, as can be seen from data in Table 1.5.

There is an evident correlation between both sets of data, but the values differ considerably. From this and other studies (37), it is evident that the tacticity index increases when polymerization temperature decreases. Martinez (38) determined the PVC dehydrochlorination rate for samples of varying tacticity index obtained by fractionation of polymers of varying polymerization temperature. Fig. 1.7 shows the relationship of both values.

TABLE 1.5.

**Tacticity index measured by IR in KBr pellets
and cast film.** (Data from Ref. 35.)

Polymerization temp., K	Tacticity index	
	KBr disc	cast film
318	1.25	1.75
258	1.50	2.20
233	1.62	2.40
213	1.60	2.55
183	2.28	3.30

Martinez explains the increase of dehydrochlorination rate, which is parallel to the increase in syndiotacticity, by easier propagation of polyénes along syndiotactic sequences as compared with that along atactic ones. His explanation is based on studies of chain scission number (by ozonolysis) for samples degraded to 0.1 and 0.3% conversion. The difference in the number of chain scissions for a sample of the lowest tacticity index (1.12) is very high, while there is no considerable change in the number of chain scissions for two other samples of tacticity index 1.25 and 1.56. Unfortunately, the data as presented cannot be used to estimate if initiation can be correlated with tacticity index because the authors (38) did not quote the number of scissions for virgin polymers. Instead, they compared only the molecular weight of initial polymer with the molecular weight of polymer which was degraded and then subjected to ozonolysis. Therefore, the final result includes the sum of double bonds initially present in polymer and those newly formed. In further studies of the same research group (39), we can find data showing that the initiation starts more easily from atactic sides than it does at syndiotactic ones. Hence, summary of these works shows the syndiotactic segments as less vulnerable to degradation initiation but at the same time more able to promote polyene propagation when it starts.

38

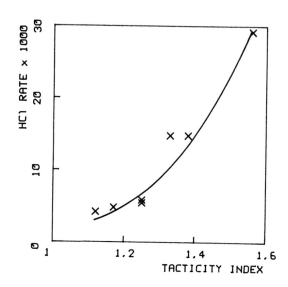

Fig.1.7. PVC dehydrochlorination rate at 453K up to 0.3% conversion versus tacticity index of polymer. (Data from Ref. 38.)

Fourier transform IR studies revealed (40) that annealing at temperatures above T_g resulted in an increase of trans-syndiotactic structures.

Raman spectroscopy seems to be the best-suited method for configuration studies. First of all, the specimens do not need to be specially prepared since material in its original state can usually be analyzed, thus diminishing the risk of configurational and conformational changes. A sample of an isotropic nature can be analyzed; therefore, the method is able to provide information about the symmetry of the species. Finally, better resolution is provided by Raman spectroscopy, and peaks can be more easily assigned to the particular structures. Table 1.6 shows the assignment of the bands to corresponding structures (41).

TABLE 1.6.

Assignment of peaks obtained by Raman spectroscopic measurements.
(Modified from Ref. 41.)

Frequency	Configuration	Structure	C-Cl type
614	syndiotactic	T T T T T T	S_{HH}
623	isotactic	T G T T G T	S_{HH}
634	isotactic	T G T G T G	S'_{HH}
647	syndiotactic	T T T G	S'_{HH}
680	syndiotactic	T T G G	S_{HC}
692	isotactic	T G T G T G	S_{HC}
704	syndiotactic	T T T G	S_{HC}

φ - Cl atom; T - trans bond; G - right handed gauche bond
(refer to explanations below); G_* - left handed gauche bond;
S_{HC} - conformation of secondary chlorine trans to hydrogen;
S_{HH}; S'_{HH} - for bent structures.

One important conclusion remains to be drawn from the studies
on configuration. Milán (42) presented the results of his
studies on dehydrochlorination of samples of atactic and
syndiotactic PVC in two different solvents; i.e.,
dimethylformamide (DMF) and hexamethylphosphortriamide (HMPT).
Fig. 1.8. shows the data obtained in both solvents. From Fig.
1.8 we can clearly see that the effect of the solvent prevailed
over the properties of the polymer. A similar conclusion can be
drawn from a comparison made between thermal degradation of a
solid sample and a polymer in solution, where the relations
obtained also showed a different course and, therefore, implied
different conclusions. This point is stressed because many
studies on PVC degradation are done in solution, and frequently
they are not correlated with similar observations for polymer
alone. As a result, we have created confusion due to
contradictory conclusions.

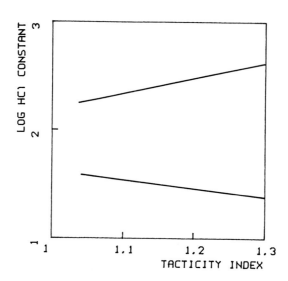

Fig.1.8. The reaction rate of dehydrochlorination versus PVC tacticity index. (Modified from Ref. 42.)

Considering that the structure of degraded polymer includes thermally induced sequences of conjugated double bonds, we may say that they would not affect configuration in the sense that the polyenes have no specific configuration. This is just opposite to the effect on conformation. Conformation is a common name for the spatial arrangement of atoms forming a molecule which, due to rotation around the single bond or bonds, may form isomers of differing properties. There are two types of conformation - one called cis-trans isomery, and the other conformations of the gauche type.

The above data on Raman peaks assignment have already quoted some conformational structures believed to be present in the PVC chain. Newman projections of the three rotational states of $CHCl-CH_2$ skeletal bonds would help us to better understand these structures.

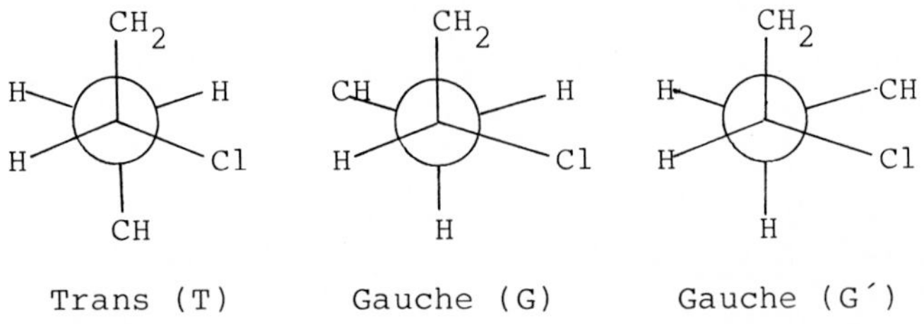

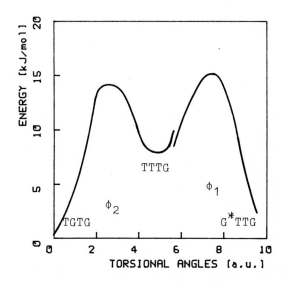

Trans (T)   Gauche (G)   Gauche (G′)

Fig.1.9. Newman projections of the three rotational states of $CHCl-CH_2$ and CH_2-CHCl skeletal bonds.

Conformational notation is, therefore, as follows: T = 180°; G = 60° and G′ = 300°. Several interactions are taken into account in the calculation of the conformational properties of vinyl polymers. These include gauche three-bond main chain interaction, the skew Cl three-bond interaction, $CH_2...CH_2$, $CH_2...Cl$ and $Cl...Cl$ four bond interaction. Details of these calculation methods are included in many research papers (43-47). Recent calculations have yielded data for the transition energy path between various conformational states (Figs. 1.10-1.12) (47).

Fig.1.10. Least energy path for transition between TGTG and G^*TTG through TTTG intermediate state. (Modified from Ref. 47.)

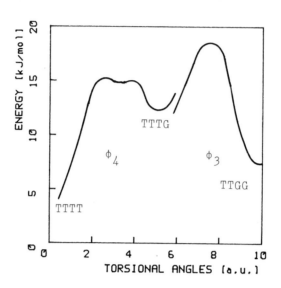

Fig.1.11. Least energy path for transition between TTTT and TTGG via TTTG. (Modified from Ref. 47.)

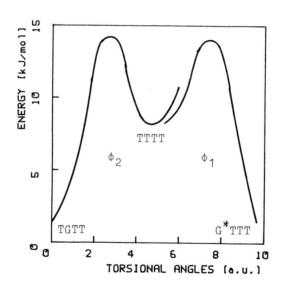

Fig.1.12. Least energy path for transition between TGTT and G*TTT via TTTT. (Modified from Ref. 47.)

The torsional angles for the figures can be seen from the diagram below:

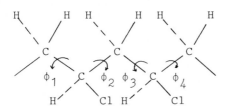

and they give a representation of changes occurring, while the energy level needed for change can be read from the Y-axis. The above diagram shows the isotactic segment of the chain.

From these explanations and examples, it is evident that the PVC backbone is not planar, as we are used to seeing it from chemical equations; a three-dimensional space is needed in order to show its real arrangement.

We should also consider another conformational structure; that is, cis-trans isomerism. This type of isomery is related to bonds that lack the freedom of rotation, such as the C=C double bond. It means that both structures, well-known from basic organic chemistry:

will be related in our case to polyene sequences.

Cis-trans isomerism has not been studied in PVC either experimentally or by molecular calculations; therefore, in order to explain this phenomenon, we should try to find related substances for which more data are available. We have two possibilities: either to consider the low molecular models having a similar number of conjugated double bonds, such as, for example, carotenoids (but in this case we shall expect the influence of an ionon ring on charge distribution, and, at the same time, on conformation changes); or to take into account a polymer having in its backbone a conjugated double bond, and here polyacetylene serves as a good example. Taking into account the

second model, we are able to say more about the polyene
structure, as it was discussed in numerous publications (48-56).
From these studies it is evident that polyacetylene exists in
four configurational forms that are well characterized by the
following structures:

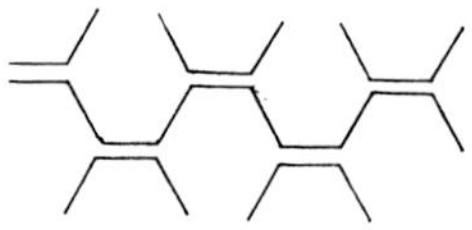

cis-transoid

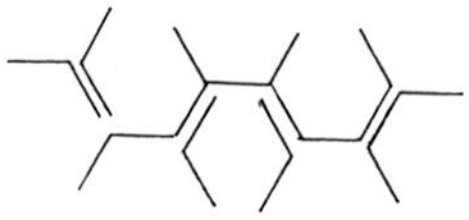

trans-cisoid

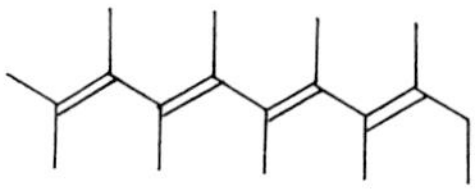

trans-transoid

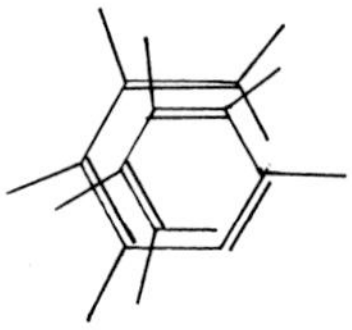

cis-cisoid

When the polymer is obtained, at considerably low temperature,
e.g., 195K, it yields primarily cis-$(CH)_x$, which is, however, a
thermodynamically unstable form that can be isomerized into
all-trans polyacetylene. The polymer initially obtained by low
temperature polymerization exists in the cis-transoid
conformation. The isomerization process can occur due to thermal
treatment, even at low temperatures; for example, it was
discovered that isomerization occurs during IR spectra
measurement, at which the temperature was estimated to be around
313K. The activation energy of the isomerization process is
initially very low (67 kJ/mol). As the isomerization process
continues, the activation energy increases to 117 kJ/mol. It is
thought that the first step in isomerization is defect-induced
(54). From the viewpoint of structural changes, gradual
isomerization is due to rotation around single bonds as shown
below:

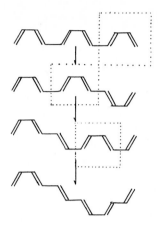

It continues until an all-trans structure is finally obtained.
The isomerization process is temperature controlled, and full
conversion takes place at 453K in just a few minutes. The
changes which take place are irreversible; the trend always
observed is from cis to trans and never the opposite. Similar
changes as far as conformation is concerned can be
oxygen-stimulated (52), but in this case the changes are more
complicated, as can be seen from electrical conductivity
measurements (Fig. 1.13)

Initially, oxygen acts as a dopant (fast action) and then as an
oxidant (slow action). This last phenomenon leads to product
destruction, producing various kinds of defects, reducing chain
length, destroying conjugation and changing the mechanical
properties of polymer.

It is interesting to notice that there is a difference in color
between both cis and trans forms. While the cis form is blue,
the trans form is red. It is therefore evident that what we
observe in PVC is evidently the trans-form. Now, we should
realize what this type of configuration means in the case of PVC.
When the initial PVC lacks a conjugated system, its conformation
(previously discussed) can be seen as either helical or lamellar
- in other words, the polymer has developed a spatial arrangement
and great freedom to change it, as was seen from the low energy
level needed to change conformers (Figs. 1.10-1.12). If during
degradation we expect the formation of cis-transoidal or
trans-cisoidal forms, the situation would not be changed much,

since they also have both the freedom to change conformation and
a three-dimensional spatial arrangement. But if, due to
isomerism, we arrive at an all-trans form, this part of the
polymer has a flat, rigid conformation, so the degraded polymer
includes in its chain fragments of a certain rigidity. This has
serious implications in technology, and the use of PVC products
as thermal treatment affects the elasticity of material; if such
material can change, the result would be still worse, as due to
oxidation the final product would be brittle.

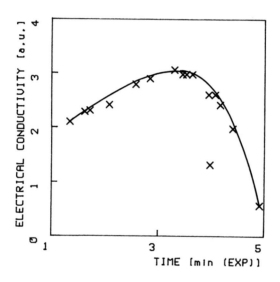

Fig.1.13. Electrical conductivity of polyacetylene versus
oxidation time. (Modified from Ref. 52.)

The reference to studies done on polyacetylenes should help us
to understand some phenomena which we were unable to determine
for PVC, and at the same time, it may stimulate an interest in
PVC studies in order to confirm observations derived from other
fields of study.

Finally, we should see how configuration and conformation may
eventually affect the thermal stability of the polymer under
discussion. We have already discussed this problem in the case
of configuration, and now we would expect to see some data on
conformation in relation to thermal degradation. One example can
be taken from polyacetylenes. It was discovered (54) that the

crosslinking of polyacetylenes due to the Diels-Adler reaction is
a stereospecific process described by the following structures:

The process can only occur if a segment with a diene and a
dienophile structure (cis-transoid and trans-cisoid) are facing
each other. It should be noticed that Diels-Adler crosslinking
breaks the conjugation. The other example of the configuration
effect on PVC degradation can be taken from Caraculacu´s work
(13), which investigated the effect of the presence of a bulky
group of three different conformations at a branching point on
the reaction probability with phenol and thiophenol.

The other example, which will be discussed in more detail with
the mechanism of thermal degradation, is connected with the
benzene formation mechanism in the course of the thermal
degradation process. This is actually a case which should still
be reconsidered in the light of results from polyacetylenes as
the cyclization so far proposed seems to be structure retarded.

Svetly (57) has also offered some structural considerations in
order to explain the effect of β-keto-unsaturated structures
which can lead to the following changes:

Other examples can still be found, and they will all lead us to
the same conclusion: that there is no doubt that configuration
and conformation play an essential role in the PVC degradation
process.

1.3.3. Chain folds

Chain folds should be discussed together with other data
concerning the crystalline structure of PVC, as they are part of

the same phenomenon. An exception was made because the results
discussed here are based on potential-energy calculations
contrary to the experimental nature of data that is included in
crystalline-state studies discussed at the end of this chapter.
The data presented on chain folds are based on the only available
publication so far (58). Some of the input data on which
calculations are based include X-ray measurement results of
syndiotactic PVC obtained from 0.1% solution in chlorobenzene
cooled from 423K (59). The bond length and angles are assumed for
the usually used values for PVC. The model is composed of two
stems, each one formed from 18 vinyl chloride monomer units
joined by a fold. The resulting models corresponding to (010)
and (400) folds are shown in Fig. 1.14.

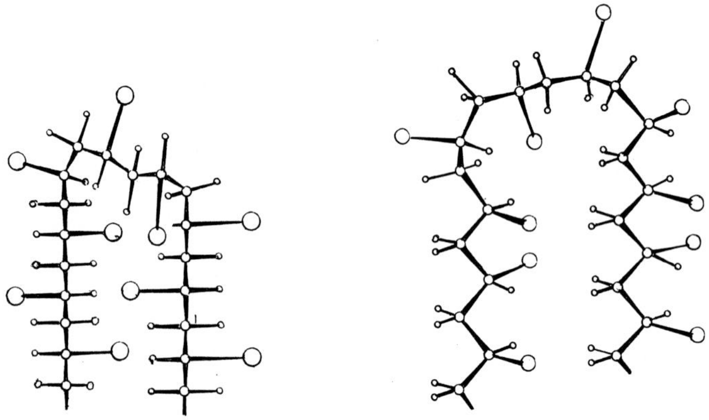

Fig.1.14. (010) and (400) folds of syndiotactic PVC. (Modified
from Ref. 58.)

It is interesting that good contact among atoms of different
stems is lacking, which is due to the length that a fold can
assume owing to its allowed conformation. Weak C-H...Cl-C
hydrogen bonds are present in the fold region. The presence of
such bonds was predicted earlier from IR studies as 64 cm^{-1} band
was recorded (60). The other interesting feature of these folds
is connected with their rather inflexible nature, which opposes
the close approach of the stems. When investigators introduced
one ethylene monomer into the fold, they obtained better crystal
packing, due to the fold flexibility.

These studies, although based on an assumption which cannot be
easily confirmed experimentally, seem to suggest other

possibilities when considering the PVC thermal degradation process. Earlier it was said that syndiotactic configuration diminishes the probability of initiation, which might eventually be explained by the chain rigidity observed in these studies. The proposed lack of chain flexibility and its uniform structure, if confirmed, should have important implications for the model of degradation. It is true that the energy needed for folding and refolding is always available during the degradation process, but at the same time the possible presence of ordered structures would always decrease the reaction probability and should contribute to better system stability. Observations from these studies could be relevant for an explanation of the contradictory facts mentioned by Braun in his review article (9). Commenting on studies on the effect of syndiotacticity on PVC thermal stability, Braun observed that from one perspective it is proved that higher syndiotacticity increases the dehydrochlorination rate as confirmed by Milán´s results, but at the same time it is known that polymer obtained in lower temperatures (more syndiotactic) is also more stable. Braun explained this contradiction in both trends by the possible influence of other factors, such as various labile structures, molecular weight and so on. Actually, Milán pointed out that only polyene propagation is faster with syndiotacticity increase, while the initiation rate is lower. These two features are regulated by different principles. In the case of propagation, the energy level plays an essential role. It is regulated by the overall charge distribution related to chemical structure and configuration as well, without reference to HCl and other catalytic effects. On the other hand, initiation probability is related (among other factors) to collision probability, which should be lower when part of the chain is in a restricted configuration state.

Summarizing the above, it is difficult to say at the moment how far these structural properties determine the degradation mechanism, since the information available is too limited, but it is important to consider the existence of structural factors, as they may prove relevant in explaining certain phenomena.

1.3.4. Chain thickness

Privalko (61), based on the following equation:

$$A = v_c M/dN_A = m_o v_c/d_o N_A$$

where:

  A - macromolecular cross-sectional area ($A = a^2$),
  a - chain thickness,
  v_c - crystalline specific volume,
  M - molecular weight of monomer unit,
  d - identity period,
  $m_o = M/n$,
  n - number of main chain bonds in mer,
  $d_o = d/n$,
  N_A - Avogadro's number,

has calculated chain cross-sectional areas for different polymers. For vinyl polymers this equation can be simplified as follows:

$$A = 1.41m_o$$

From these calculations, we have a value for PVC of 2.718 nm, and for polyacetylene, 1.828 nm. It is evident that these are values describing only the thickness of the polymer chain at one point; that is, they would not correspond to values obtained from either crystallographic measurement or conformational calculations. By conformational calculations Conte (58) has arrived at a lamellar thickness of 6.0 nm for PVC. From the small angle diffraction, the X-ray long spacing for PVC also equals 6.16 nm (62). For cis-polyacetylenes the lamellar thickness is estimated to be in the range of 5-10 nm, while for all-trans conformation it should be as estimated by Privalko (61).

1.4. The spatial arrangement of polymer chains
Crystalline and amorphous fractions

The structural studies, referred to here, should, at best, summarize the above discussion on stereochemical properties of poly(vinyl chloride) and give an idea as to how polymer chains are distributed so far as spatial arrangement is concerned. It is, of course, not possible that the following discussion can be fully detailed because the material under study poses many complications since it is not crystalline. It is only recently that any reasonable information can be found in the literature. About ten years ago, if we had been able to look at data available from our present understanding, this sub-chapter would be almost completed.

Structural studies are usually associated with results obtained by the use of an X-ray diffractometer, while in the case of PVC they contribute to only a fraction of the material available. Natta (63) has established the unit-cell dimensions as follows: a=1,04 nm, b=0.53 nm and c=0.514 nm. The space group is Pcam, Z=4. These data have been confirmed more recently by Wilkes (62). These studies concern relatively highly crystalline material of syndiotactic configuration. Similar data for industrial polymers cannot be obtained due to their low crystallinity. Practically, three main methods are applied in order to measure the crystalline nature of polymer, i.e., IR spectroscopy, X-ray diffraction and density determination.

Chartoff, et al (64) were interested in the possibilities of characterizing the internal order in PVC material by application of IR spectroscopy. This technique is frequently used, but the method of interpreting results is still under dispute. First of all, one should take into consideration that industrial polymers differ in their IR spectrum, particularly in the carbon-chlorine stretching region, where three bands are located at 610, 635 and 690 cm^{-1}. Not only the crystallinity of the sample affects IR absorption, but also the physical state, tacticity, history of the specimen, and so on. The usual approach taken in IR studies is to compare the absorption intensity at the crystalline active frequency with that of the crystalline inactive band. The last band acts as a sort of internal standard. In early studies, absorption at 690 cm^{-1} was used as the internal standard, but it was soon discovered that this band is closely related to tacticity, and therefore, it cannot be regarded as a good internal standard, since crystallinity and syndiotacticity are interrelated. The other possibilities were therefore exploited so as to use either absorption at 2920 or 2960 cm^{-1}, which are

$C-H_2$ antisymmetric stretching mode vibrations and C-H stretching mode, respectively, and the previously mentioned band at 610 cm^{-1}. Considering the crystalline active absorbance, it seems that the 635 cm^{-1} frequency would be the most appropriate. From Chartoff's studies on spectral changes during annealing, it is evident that the ratio between A_{635}/A_{610} serves the purpose of crystallinity value determination (64).

Density measurement is also a crystallinity indicator as it depends on the additive contribution of amorphose and crystalline phases. The average crystallinity value (X_c) can be calculated from the following formula:

$$X_c = \frac{d_c d_a /(d - d_c)}{d_a - d_c}$$

where:

d_c - crystalline phase density,

d_a - amorphous phase density (determined in amorphous condition of sample),

d - density of particular sample.

The density measurement is less sensitive than IR studies, and both methods suffer from difficulties in obtaining a standard of amorphous PVC. This problem would be better understood if we were to discuss it based on X-ray data. Assessing the crystallinity of polymer by wide-angle X-ray scattering measurement, we assume the two-phase concept of the polymer structure, i.e., we imply that there is a difference between the crystalline area and the amorphous area in the polymer sample and on a diffractogram. Therefore, the result of the crystallinity index depends greatly on how we choose the amorphous area, or in practice, the amorphous sample. It is a common practice to apply a quenched sample as an amorphous standard, but this sample has a profile similar to a crystalline diffraction pattern with diminished intensity by, let's say, the small size of its reflecting crystallites. It will be interesting to quote in full the Guerrero's comments (65) on this subject:

> Under these circumstances it is difficult to surmise what the true shape of amorphous scattering in PVC will be. There are serious doubts as to the validity of the

widely accepted quenched PVC amorphous profile. In any case, the presence of the intermediate state of order would certainly question even further the use of a simple two-phase concept to derive crystallinity values from PVC diffractometer curves.

These difficulties were experienced by this author (66) in studies carried out about ten years ago, in which the crystallinity index was measured for about 20 samples of industrial polymers in order to use it as a factor characterizing their processability. The results of X-ray measurements, correlated with the polymers´ thermal and rheological properties, resulted in a relationship of low correlation that was due to the method of amorphic fraction assessment. Based on the above discussion, one may understand why in recent papers (64, 65) the former estimation of the crystallinity value of PVC as 10% or below is not held to be adequate due to an incorrectly-chosen amorphous fraction. It is suggested that the crystallinity of PVC should rather be in the range of 10-30% (64).

Figs. 1.15 and 1.16 illustrate the three above-mentioned methods.

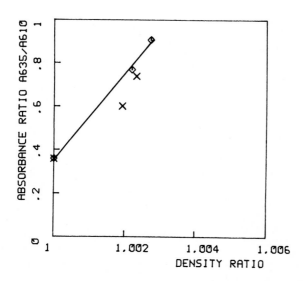

Fig.1.15. Correlation between the IR and density crystallinity indices. (Modified from Ref. 64.)

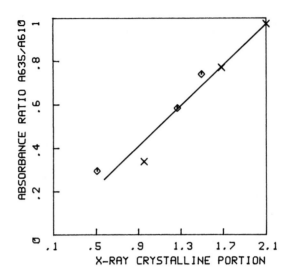

Fig.1.16. Correlation between the IR and X-ray crystallinity indices. (Modified from Ref. 64.)

These data can also illustrate the importance of reference (amorphous) state selection. As long as this problem is not solved, the data should be regarded as approximations and compared only for particular conditions. All three methods allow analysis of crystalline morphology in internally comparable units, but IR is sensitive to molecular ordering in the smaller scale (67).

Two types of crystals can be found in PVC - the lamellar type with foldings of width estimated to be in the range of 50-300 nm, and the fringed micells type which are characterized by the diagram below:

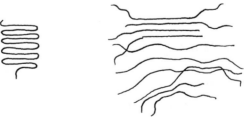

Fig. 1.17. PVC crystalline structures.

The existence of fringed micells requires segregation of long syndiotactic sequences. The entangled chains protect crystalline domains against plastification, which was observed microscopically (67). Crystallites in industrial PVC should be kept small or imperfect for the above reason and also because the melting point of crystalline, highly-syndiotactic PVC is higher than the thermal stability of polymer (68). This is well-known in practice from the production of highly transparent foil, wherein all the unplasticized or unmelted polymer particles affect product quality, and therefore, this process needs special polymer selection. Additionally, polymer melt is passed through the filter to eliminate unmelted material.

The existence of folded lamellar structures was also confirmed by differential scanning calorimetry measurements (69). It was found that the melting-point of extended-chain lamella was restricted by the stem length degree of polymerization. The stem length depends on the degree of polymer syndiotacticity and is estimated in the range of about 12-16 monomer units.

Summarizing the above discussion on structural properties, we can see that we are still far behind expectations, yet we observe considerable improvement in the methods of study. It must be emphasized that various structural calculation methods are applied to PVC, and these results are definitely stimulating, since there is a common ground for comparison. Experimental studies alone are unable to cope with the variety of parameters involved; thus best hope for the future advances is associated with a new approach to problem solution, including interchanging ideas from computational techniques and experimental observations. It is still an open question on how to improve the methodology of experimental methods, as this factor is presently determining the efficient use of mathematical modelling in polymer stereochemistry.

REFERENCES

1. R.D. Levine and R.B. Bernstein in **Molecular Rection Dynamics**, Oxford University Press, New York, 1974.
2. A. Maccoll and P.J. Thomas, **Nature**, 1976(1955)392.
3. K.A. Holbrook, R.W. Walker and W.R. Watson, **J. Chem. Soc., B**, 577(1971).
4. M.D. Joesten and I.J. Schaad in **Hydrogen Bonding**, Marcel Dekker, New York, 1974.

5. R. Petiand and Quand-Tho-Pham, **Makromol. Chem.**, 178(1977)741.
6. U. Schwenk, I. Koenig, F. Cavanga and B. Wrackmeyer, **Angew. Makromol. Chem.**, 83(1979)183.
7. A. Caraculacu and E. Bezdadea, **J. Poly. Sci., Polym. Chem. Ed.**, 15(1977)611.
8. T. Hjertberg and E.M. Sörvik, **J. Macromol. Sci.-Chem.**, A17(1982)983.
9. D. Braun, **Pure Appl. Chem.**, 53(1981)549.
10. E.C. Buruianǎ, G. Robilǎ, E.C. Bezdadea, V.T. Barbinta and A.A. Caraculacu, **Eur. Polym. J.**, 13(1977)159.
11. K.S. Minsker, V.V. Lisitsky, S.V. Kolesov and G.E. Zaikov, **J. Macromol. Sci.-Rev. Macromol. Chem.**, C20(1981)243.
12. K.S. Minsker, V.V. Lisitsky and G.E. Zaikov, **Vyssokomol. Soed.**, 23(1981)483.
13. A.A. Caraculacu, **Pure Appl. Chem.**, 53(1981)385.
14. A.J. de Vries, C. Bonnebat and M. Carrega, **Pure Appl. Chem.**, 25(1971)209.
15. J. Lyngaae-Jørgensen, **J. Chromatogr. Sci.**, 9(1971)331.
16. K.B. Andersson, A. Holmstrom and E. Sörvik, **Makromol. Chem.**, 166(1973)247.
17. A. Rigo, G. Palma and G. Talamini, **Makromol. Chem.**, 153(1972)219.
18. G. Park, **J. Macromol. Sci.-Phys.**, B14(1977)151.
19. W.H. Starnes, **Dev. Polym. Deg.**, 3(1981)135.
20. W.H. Starnes, F.C. Schilling, K.B. Abbàs, R.E. Cais and F.A. Bovey, **Macromolecules**, 12(1979)556.
21. W.H. Starnes, F.C. Schilling, I.M. Piltz, R.E. Cais, J.D. Freed and F.A. Bovey, **Third Int. Symp.**, Cleveland, (1980)58.
22. W.H. Starnes, F.C. Schilling, I.M. Plitz, R.E. Cais and F.A. Bovey, **Polym. Bull.**, 4(1981)555.
23. T.N. Bowmer, S.Y. Ho, J.H. O´Donnell, G.S. Park and M. Sallem, **Eur. Polym. J.**, 18(1982)61.
24. J. Bauer and A. Sabel, **Angew. Makromol. Chem.**, 47(1975)15.
25. M.H. George and A. Garton, **J. Macromol. Sci.-Chem.**, A11(1977)1389.
26. K. Mitani, T. Ogata, H. Awaya and Y. Tomari, **J. Polym. Sci.**, A-1, 13(1975)2813.
27. S. Crawley and I.C. McNeill, **J. Polym. Sci., Polym. Chem. Ed.**, 16(1978)2593.

28. A. Crosato-Arlandi, G. Palma, E. Peggion abd G. Talamini, **J. Appl. Polym. Sci.**, 8(1964)747.

29. G. Martinez and J. Millán, **Angew. Makromol. Chem.**, 75(1979)215.

30. G. Robilă, E.C. Buruianǎ and A.A. Caraculacu, **Eur. Polym. J.**, 13(1977)21.

31. A.E. Hamielec, R. Gomez-Vaillard and F.L. Marten, **J. Macromol. Sci.-Chem.**, A17(1982)1005.

32. W.W. Graessley, W.C. Uy and A. Ghandi, **Ind. Eng. Chem., Fundam.**, 8(1969)696.

33. T. Kelen, **J. Macromol. Sci.-Chem.**, A12(1978)349.

34. E.M. Sörvik, **J. Appl. Polym. Sci.**, 21(1977)2769.

35. M.E. Carrega, **Pure. Appl. Chem.**, 49(1977)569.

36. Q.T. Pham, J. Millán and E.L. Madruga, **Makromol. Chem.**, 175(1974)945.

37. J. Millán, C. Mijangos, J. Martinez, P. Berticat and J. Chauchard, **Eur. Polym. J.**, 15(1979)615.

38. G. Martinez and J. Millán, **J. Macromol. Sci.-Chem.**, A12(1978)489.

39. J. Millán, G. Martinez and C. Mijangos, **J. Polym. Sci., Polym. Chem. Ed.**, 18(1980)505.

40. J.L. Koenig and M.A. Antoon, **Appl. Optics**, 17(1978)1374.

41. M.E.R. Robinson, D.I. Bower and W.F. Maddams, **Polymer**, 19(1978)773.

42. J. Millán, **J. Macromol. Sci.-Chem.**, A12(1978)315.

43. P.J. Flory and A.D. Williams, **J. Am. Chem. Soc.**, 91(1969)3118.

44. J.E. Mark, **J. Chem. Phys.**, 56(1972)451.

45. J.E. Mark, **J. Am. Chem. Soc.**, 94(1972)6645.

46. F. Cantera, E. Riande, J.P. Almendro and E. Saiz, **Macromolecules**, 14(1981)138.

47. R.H. Boyd and L. Kesner, **J. Polym. Sci., Polym. Phys. Ed.**, 19(1981)375.

48. L. Terlemezyan and M. Mihailov, **Makromol. Chem., Rapid Commun.**, 3(1982)613.

49. T. Yamabe, K. Akagi, H. Shirakawa, K. Ohzeki and K. Fukui, **Chem. Scr.**, 17(1981)157.

50. P. Bernier, F. Schue, J. Sledz, M. Rolland and L. Giral, **Chem. Scr.**, 17(1981)151.

51. T. Yamabe, K. Akagi, K. Ohzeki and K. Fukui, J. Phys. Chem. Solids, 43(1982)577.

52. M. Aldissi, M. Rolland and F. Scue, Phys. Stat. Sol., 69(1982)733.

53. A. Montaner, M. Galtier, C. Benoit and M. Aldissi, Solid State Commun., 39(1981)99.

54. H. Kuzmany, E.A. Imhoff, D.B. Fitchen and A. Sarhangi, Mol. Cryst. Liq. Cryst., 77(1981)197.

55. T. Yamabe, T. Matsui, K. Akagi, K. Ohzeki and H. Shirakawa, Mol. Cryst. Liq. Cryst., 83(1982)125.

56. C.I. Simonescu and V. Percec, Prog. Polym. Sci., 8(1982)133.

57. J. Světlý, R. Lukáš, J. Michalcová and M. Kolinský, Makromol. Chem., Rapid Commun., 1(1980)247.

58. G. Conte, L. D´Ilario, N.V. Pavel and E. Giglio, J. Polym. Sci., Polym. Phys. Ed., 17(1979)753.

59. A. Nakajima and S. Hayashi, Kolloid Z. Z. Polym., 229(1969)12.

60. A.V.R. Warrier and S. Krimm, Macromolecules, 3(1970)709.

61. V.P. Privalko, Macromolecules, 13(1980)370.

62. C.E. Wilkes, V.L. Folt and S. Krimm, Macromolecules, 6(1973)235.

63. G.Natta, I.W. Bassi and P. Corradini, Atti Accad. Naz. Lincei Cl. Sci. Fis. Mat. Nat. Rend., 31(1961)17.

64. R.P. Chartoff, T.S.K. Lo, E.R. Harrell and R.J. Roe, J. Macromol. Sci.-Phys., B20(1981)287.

65. S.J. Guerrero, D. Meader and A. Keller, J. Macromol. Sci.-Phys., B20(1981)185.

66. T.J. Bartczak, Z. Gałdecki, A. Wypych and J. Wypych, unpublished studies.

67. R. Biais, C. Geny, C. Mordini and M. Carrega, Brit. Polym. J., (1980)179.

68. A. Michel and A. Guyot, J. Polym. Sci., Part C, 33(1971)75.

69. J.N. Hay, F. Biddlestone and N. Walker, Polymer, 21(1980)985.

CHAPTER 2

THE CHEMICAL ASPECTS OF THE PVC THERMAL DEGRADATION PROCESS

2.1. The reasons for polymer instability

The comparison between the thermal stability of PVC and that of some low molecular weight substances having a chemical structure similar to PVC can be seen from the data collected in Table 2.1.

TABLE 2.1.

The activation energy and degradation temperature
of some organic substances. (Data from Ref. 1 and 2.)

Compound	Degradation temp., K	Activation energy, kJ/mol
1,4,7-trichloropentane	503-533	95
1,4,9-trichlorononane	493-523	132
8-chlorohexadecane	509-557	151

If one compares this temperature range with the usual temperature of PVC degradation, it seems reasonable to suggest that PVC contains some additional structures, atypical for polymer, which cause the degradation temperature to decrease. This assumption underlies the present theory of PVC degradation. It is suggested, therefore, that PVC molecules contain structures that are more easily destroyed on heating and that are responsible for lowering the temperature of PVC decomposition. This hypothesis makes sense when compared with data for some substances of low molecular weight shown in Table 2.2.

One can see that the double bond in the vicinity of chlorine has an essential effect on the degradation temperature, even when the activation energy is unchanged. According to our present understanding, the chloroallyl groups are regarded as the most

likely center for the dehydrochlorination reaction. The first
essential studies were done by Frye (3-5) and Bengough (6), who
substituted allylic chlorine atoms by carboxylic acid rests and
found adequate changes in the IR spectrum.

TABLE 2.2.

The activation energy and degradation temperature
of some organic substances. (Data from Ref. 2.)

Compound	Degradation temp., K	Activation energy, kJ/mol
4-chlorododecene-2	430-453	95
4-chlorohexene-2	433-463	57

Recently, Buruiana (7) presented the results of thermal
degradation of samples before and after bromination, which are
given by Fig. 2.1.

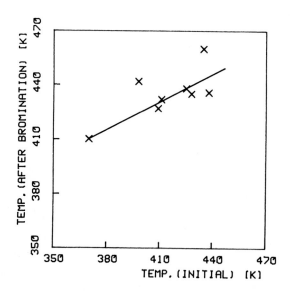

Fig.2.1. Degradation temperature of initial PVC plotted against
that of brominated PVC. (Data from Ref. 7.)

One can see that bromination has increased the sample´s thermal
degradation temperature by roughly 20-30K. Presently, almost

every investigator agrees that the chloroallylic group is the most important for the initiation of dehydrochlorination.

Most recently, Iván (8) studied the effect of allylic chlorine introduced into the initial polymer by controlled chemical dehydrochlorination. His results, shown in Fig. 2.2, are in agreement with other results referred to earlier.

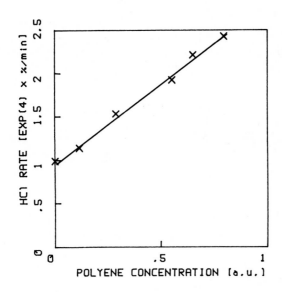

Fig.2.2. Initial rate of HCl loss versus the number of internal unsaturations $S=M_n/(M_n-1)$. (Modified from Ref.8.)

TABLE 2.3.

Characteristics of samples from **Fig. 2.2.** (Modified from Ref. 8.)

	Sample		
	1	2	3
Monomer conversion, percent	75	82.5	85
M_n	35,900	26,000	35,100
M_w	75,000	82,900	82,700
Total double bonds/1000 mers	1.19	1.93	1.93
Internal unsaturations/1000 mers	0.16	0.44	0.77
Labile chlorines/1000 mers	1.50	2.54	2.15

The strong effect of this group can be seen from Fig. 2.3.

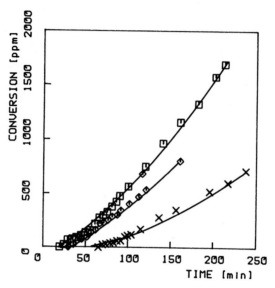

Fig.2.3. PVC thermal degradation in solution at 448K. PVC samples are discussed in Table 2.3. × - 1, ◊ - 2, ☐ - 3. (Modified from Ref. 9.)

Internal unsaturations are the only values that correlate with the dehydrochlorination rate, which may suggest that the chloroallylic groups have the strongest effect on the initiation rate. Similarly, total labile chlorine determined by phenolysis correlates with the PVC dehydrochlorination rate (Fig. 2.4).

Here the relationship is more complicated, although we have a very good correlation in every group, which can be understood since in the case of each group, total labile chlorine gives the total number of chlorine atoms bound to carbon at the branch point and in the vicinity of the double bond, but the ratio of concentration of both chlorine types is different. The tertiary carbon atoms are either bound to hydrogen (more frequently) or to chlorine at long branch points. The question is, do they influence the initiation rate of dehydrochlorination? From studies of low-molecular substances, it is known that they should, but so far no one has proved it. In Buruiană´s paper (7) data are available on both the concentration of tertiary chlorine and the degradation temperature of eight PVC samples. For extreme concentrations of tertiary carbon bound chlorine of 0.18 and 1.17 per 1000 monomer units, the degradation points are 427 and 410K, respectively, showing that these groups may affect PVC thermal stability. At the same time, among the samples analyzed, two of 0.18 tertiary chlorines per 1000 monomer units have a degradation point at 427 and 460K, respectively, which means a difference even greater than that caused by the eventual concentration of tertiary chlorine. Table 2.4 shows the correlation factors between PVC thermal stability and allylic chlorine and tertiary chlorine based on data contained in papers by Buruianǎ (7) and Hjertberg (10).

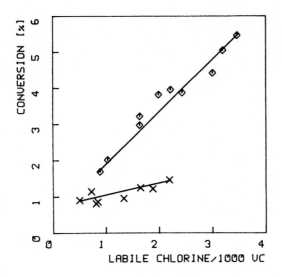

Fig.2.4. PVC dehydrochlorination rate versus the amount of labile chlorine. × - industrial, suspension PVC, ◇ - PVC obtained under subsaturation conditions. (Modified from Ref. 10.)

In the case of Buruianǎ´s data (7), the dehydrochlorination temperature was correlated with the total labile chlorine (phenolysis) and tertiary chlorine (the difference between phenolysis before and after bromination). Data from Hjertberg´s work correlated in the above table include the dehydrochlorination rate and allylic chlorine (ozonolysis and molecular weight change determination). Table 2.4 confirms that the data on labile chlorine show hardly any correlation with the dehydrochlorination rate. The results demonstrate how difficult it is even to say if tertiary carbon-bound chlorine affects the thermal dehydrochlorination rate. The ambiguity of tertiary carbon-bound chlorines effect is not necessarily the result of imprecise data, but rather a result of the principle of data use; we are trying to correlate just one factor capable of affecting dehydrochlorination, whereas these factors are probably numerous.

TABLE 2.4.

The correlation factor (r^2) between PVC thermal stability and allylic chlorine and tertiary carbon-bound chlorine.
(Data from Refs. 7, 10.)

Reference	Correlation factor for:		
	allylic Cl	tertiary Cl	total labile Cl
7	–	0.53	0.31
10(suspension PVC)	0.41	–	0.92
10(subsaturated PVC)	0.81	–	0.82

The oxygen-containing groups' effect on the PVC dehydrochlorination rate is even more complicated since more information is available, but the results are even more controversial. Braun (11), who studied in detail the effect of oxygen on polymerization and polymerizate, and also the copolymerization of vinyl chloride with CO, concluded his work as follows:

> thermal degradation tests reveal a close correlation between dehydrochlorination rate and the number of internal double bonds. Incorporated carbonyl groups, even at up to 2 mol%, affect only slightly the thermal stability of PVC. Residual peroxides, however, increase remarkably the initial dehydrochlorination rate.

Conversely, anybody who works on PVC degradation in air would show the PVC dehydrochlorination rate to be a function of oxidation degree, as, for instance, Kelen (12) does, whose results are given in Fig. 2.5.

Kelen's results (12) include the PVC dehydrochlorination rate

for samples previously oxidized at varying durations. Similar
data can be found in many other papers discussed below. From
Fig. 2.5 one can see that the dehydrochlorination rate depends on
the amount of oxygen incorporated into the polymer. The question
is whether the resulting difference is due to experimental errors
or to the type of bonds formed by oxygen. Unfortunately, we are
still unable to identify the type of bonding in each case, which
would be crucial, especially because controversy still exists
over how to assess the initiation by β-chloroallylic groups.
Most research indicates that this group is the main site of
initiation of PVC thermal dehydrochlorination.

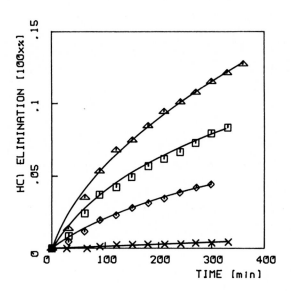

Fig.2.5. HCl emission kinetics in relation to preoxidation time
at 383K. ×- control, ◇ - 30 min, □ - 60 min, △- 180 min.
(Modified from Ref. 12.)

Minsker (13-17) proposed that internal unsaturated groups are
rapidly oxidized by ambient oxygen to carbonyl-chloroallylic
groups:

$$-\overset{\text{O}}{\underset{}{\text{C}}}-CH=CH-CHCl-$$

He developed a technique for measuring the carbonyl-chloroallylic groups, and measured many industrial and laboratory obtained polymers, showing (16) that the number of internal double bonds is exactly equal to the number of carbonyl-chloroallylic groups; in other words, he contended that no β-chloroallyl group is present under practical conditions in the polymer. Minsker has additional arguments which support his thesis:

1. The dehydrochlorination rate of β-chloroallyl low-molecular weight models is lower than that expected from the group responsible for initiation (2, 18).
2. The rate of PVC dehydrochlorination after polyenes were formed is about two to three times higher than that of β-chloroallyl models (18, 19).
3. Polymer samples prepared in the absence of oxygen, when left in air for varying times, showed a continuous growth of degradation rate until they reached the level typical for the number of internal double bonds present (13, 20).
4. The rate constant of HCl elimination accords with the rate constant exhibited by low-molecular models having two conjugated double bonds such as 7-chlorononadiene-3,5 and 6-chlorooctadiene-2,4 (16).

Based on the above, Minsker proposes the following reaction for oxidation of β-chloroallyl groups:

$$-CH_2-CH=CH-CHCl- + O_2 \longrightarrow -C(O)-CH=CH-CHCl-$$

In view of these observations, Minsker claims that PVC containing the keto-allylic group is very unstable since it has two bonds in conjugation. Here we leave the discussion of the effect of conjugation on PVC dehydrochlorination as we shall discuss it in full later in this chapter and concentrate instead on other arguments made by Minsker.

First of all, it should be understood that the results for the model substances cannot be taken as evidence in this discussion, as they are obtained under conditions entirely different from those in which PVC is degraded. This problem is well reflected in Guyot's paper (21), which shows how changing the solvent used in such a study may completely change the reaction kinetics and even its direction. Depending on the formulation under which we are carrying out experimental studies, we can find supportive arguments for any mechanism we want to propose. For instance, when 4-chlorohex-2-ene was degraded in dichloroethane, the reaction led to hexadiene in complex with HCl. Such reaction did not occur when dichloroethane was replaced by tetrahydrofuran. In some solvents, dehydrochlorination rate depends on the dielectric constant of the solvent (22). If such effects occur, we cannot seriously compare the rate of dehydrochlorination of the model substance with that of PVC and draw far-reaching conclusions.

The other perplexing question is why a carbon atom adjacent to a double bond should be so vulnerable to oxidation. The IR studies, still a matter of controversy due to the difficulties in assessing the bands, seem to suggest that the cabonyl groups present are not in conjugation with double bonds but rather are related to copolymerization with CO (23, 24).

Finally, we should raise one other question connected with the methods applied by Minsker for measuring the carbonyl-chloroallylic groups. According to him, alkaline hydrolysis occurs only at the carbonyl-chloroallylic structure, while β-chloroallylic groups are stable during this treatment. Minsker´s conclusion on the effect of the carbonyl-chloroallylic group is based on a comparison of their concentration and the concentration of the internal double bonds, which is determined by chain scission due to ozonolysis. It is possible that the difference between both results equals zero because both of them determine the same chemical moiety.

It seems that the dispute is more related to the form of Minsker´s statement than to his reasoning. Some authors apparently think that Minsker entirely excludes the possibility of the effect of other groups on the initiation rate, whereas, in fact, he tries to discuss the principles of initiation by the carbonyl-chloroallyl group (see below). Clearly, Minsker´s theory seems to exaggerate the importance of that group in certain places because he is attempting to explain even the crosslinking phenomenon by the presence of the carbonyl group, whereas rather the reverse trend is observed when one compares degradation in inert atmosphere and in the presence of oxygen.

The other problem with Minsker´s theory concerns energy since we know that oxidation processes usually need a higher energy level than that available at room temperature. This discussion will remain hypothetical until a method is found for measuring the amount of oxygen incorporated into the PVC molecule during storage under normal conditions.

The effect of head-to-head structures on PVC thermal stability was discussed by Crawley (25). Fig. 2.6 shows the comparison between experimental head-to-head PVC obtained by chlorination of polybutadiene and a purified commercial PVC sample.

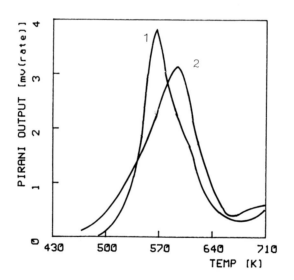

Fig.2.6. TVA curves for head-to-head (1) and head-to-tail (2) PVC samples. Heating rate 10K/min. (Modified from Ref. 25.)

Both polymers have different degradation patterns, but the thermal stability of the head-to-head sample does not seem to be worse than that of the commercial polymer. If one result so far produced were sufficient to conclude the influence of such groups on the initiation rate, we would view this group as being equivalently stable in comparison to a normal segment of a chain.

Considering the end groups´ effect on PVC instability, it is thought that both structures,

$$-CH-CH_2Cl \qquad\qquad and \qquad\qquad -CH=CH-CH_2Cl$$
$$|$$
$$Cl$$

which are detected in the polymer (10), are not the cause of polymer degradation initiation. However, research done to date is insufficient to draw any final conclusion. Presently end groups have scarcely been identified, let alone comparative studies of their effect on thermal degradation having been made.

Probably the greatest weakness in studies of PVC degradation is our inability to delineate the structures responsible for PVC degradation initiation. The chloroallylic group is likely the most labile structure of the polymer which could trigger initiation of the degradation process. The participation of any other group in degradation initiation has yet to be proved.

2.2. Kinetic stages of the process

Let us now discuss the main features of models able to characterize the kinetics of PVC dehydrochlorination. An interesting approach to PVC degradation kinetics is given by Simon (26) in his statistical model of PVC dehydrochlorination in inert atmosphere. This model is based on the energy necessary to split off the HCl molecule from the fragment of the chain which was able to acquire this energy level. The polymer is thought to be formed from N monomer units connected in one long chain. The probability of a degradation reaction is the same for all monomer units, and the energy level needed for conversion can be attained randomly. This would be true only at the beginning of the degradation process. This system allows for allyl catalysis for double bonds formed in the course of reaction. Since the activation energy is randomly distributed, activations of some units may prove ineffective, since, for instance, a unit may have already lost the HCl molecule, or a sufficient amount of energy may still not produce effective conversion because of dispersal or other circumstances.

Simon's model monitors the average length of the conjugated sequence, m, and the number of growing sequences, z, while the number of reaction steps is denoted by n. Every unitary reaction step can either increase or decrease the number of sequences z: increase by the formation of a new initiation site; decrease because two zips were separated by one unreacted monomer unit that after reaction, has combined two sequences to form one longer one. The zips number is given by the equation:

$$z = 1 + \sum_{i=2}^{n} q_i - \sum_{i=3}^{n} r_i$$

where:

q_i, r_i - probabilities of increase and decrease of zips number, respectively.

The unitary reaction step may also be effective in increasing m, i.e., average length of polyene.

The number of growing zips is given by the kinetic equation:

$$[q_z] = k_i(1 - x)/k_t$$

where:

$[g_z]$ - concentration of growing zips,
k_i - initiation rate constant,
k_t - termination rate constant,
x - dehydrochlorination degree.

The dehydrochlorination rate constant is given by the following equation:

$$k_{eff} = k_i(1 + k_p/k_t)$$

where:

k_p - propagation rate constant.

Šimon and Valko (26) tested this by using the experimental data on PVC thermal degradation from three research papers and calculating the activation energy of dehydrochlorination for several different PVC types, which was found to be in the range of 166.6. to 174.0 kJ/mol. They concluded that since the activation energy for different polymers is in agreement, HCl elimination is a characteristic property of monomer, not chain

structure. The individual polymers differ only in average zip
length. The zip length, m, also calculated for different
temperatures, was decreased with temperature increase. Based on
these model calculations, the authors also explained the
difference between the character of low molecular weight model
and PVC degradation. In the case of models, the reaction is of
an elementary type, while in the case of PVC it is a sequence of
consecutive reactions. Simon´s model does not include the
participation and effect of labile structures in PVC
degradation.

The model proposed by Danforth (27) uses even more
simplifications. The dehydrochlorination rate is calculated from
the equation:

$$\alpha = A - B - C + D$$

where A represents zipper dehydrochlorination if zips were to
produce HCl indefinitely; B substracts zips due to premature
termination; C substracts terminations and D corrects for values
due to the already-substracted term B. All terms are expressed by
three rate constants, which are the parameters being varied.
Danforth (27) also neglected the possible effect of structural
irregularities but included the effect of HCl catalysis, which is
viewed as the reason for new initiations and polyene sequence
propagation. Here again we have an entirely new approach to PVC
thermal dehydrochlorination.

Minsker (16) did not show kinetic equations, which he usually
uses for the interpretation of his results, in the form of a
model, but one can use his set of equations for the explanation
of the problem under discussion. According to Minsker, PVC
dehydrochlorination is a multistage process involving at least
four reactions occurring in parallel and/or in sequence:

1. Statistical HCl elimination from normal PVC units with
   formation of β-chloroallyl group.
2. HCl elimination due to the initiation effect of internally
   existing carbonyl-chloroallylic groups with formation of a
   system of conjugated double bonds.
3. HCl elimination due to the initiation effect of β-chloroallyl
   group, also with formation of a system of conjugated double

bonds (process viewed as being slow).
4. Polyene propagation with constant rate of HCl elimination.

The process is controlled by the concentration of isolated and conjugated double bonds, rate constants of the above reactions and constant of termination rate, time and initial HCl content in PVC. We can see that Minsker does not reject the idea that the process can be catalyzed by β-chloroallyl groups; he only insists that the reaction would be slow, and according to his determination, these groups do not exist in initial polymer as they would already have been oxidized. Actually there is a certain similarity between the views presented by Minsker (16) and Simon (26) since both believe that initiation can be started by random elimination of HCl from normal polymer units. The difference is that Minsker has introduced an additional effect, i.e. labile structure which starts the process right from the beginning. Both authors do not include the effect of HCl; in Minsker´s case it because he thinks he can avoid this effect by choosing samples of proper thickness.

At this point we arrive at a fundamental question. We said before that the basic principle of PVC thermal degradation rests on the fact that there are labile structures in PVC which can cause the initiation process, and then every other act of elimination is easier, due to the removal of HCl from the unit in the vicinity of the double bond. This process would be an unquestionable course of events if polyenes could be propagated infinitely or if there were experimental proof that the dehydrochlorination rate decreases with degradation time. But it will be shown below that neither of the two is correct. Hence, we must ask why the dehydrochlorination rate remains unchanged when the number of active polyenes in HCl emission decreases due to their termination. We already have two answers. Danforth seems to suggest that it is due to the presence of HCl, which can catalyze dehydrochlorination, while Minsker (16) and Šimon (26) contend that it is due to the random elimination which can form new active sites of polyene propagation.

Světlý (28, 29) has analyzed the model, including the participation of a mechanism of random elimination. The overall rate of HCl elimination according to his concept is expressed by the equation:

$$d[HCl]/dt = k_s + k_p x = k_s(1 + k_p/k_t)$$

where:

  k_s - HCl elimination rate constant from regular units,
  k_p - polyene propagation rate constant,
  k_t - termination rate constant,
  x - molar fraction of active sites,
  t - time.

Since the dehydrochlorination rate depends on three constants, as in the above equation, there should be no difference expected between polymer types, which is observed experimentally. Světlý (28) also concludes that the double bonds and structural irregularities do not play any essential role in PVC dehydrochlorination. This last conclusion does not seem to have sufficient support in the above model since these structures were simply not reflected there.

The model implemented with a migrating active center includes the following relationship:

$$d[HCl]/dt = k_s(1 + k_p/k_t) + k_a k_p q/k_t$$

where:

  k_a - rate constant of reaction at which the transfer of activating groups takes place,
  q - molar fraction of active groups able to migrate.

The dehydrochlorination rate is proportional to the concentration of the active group able to migrate. Based on the above models, Světlý (28) concluded:

> The observed constant rate of dehydrochlorination of PVC is a consequence of the constant concentration of active sites in the chain elimination of hydrogen chloride. This constant concentration is maintained by a dynamic equilibrium between the initiation and termination of propagating polyene sequences. The rate of initiation can be explained neither by random dehydrochlorination of regular structural units nor by usually contemplated structural irregularities, such as e.g. double bonds, branching, structural head-to-head units and initiator residues. It is likely that the initiation of propagating polyene sequences occurs via migration of activating groups between adjacent polymer chains. It seems that these activating groups are represented by oxygen containing structures (29).

Světlý's hypothesis was cited in full since, if it proves correct, it should entirely alter the direction of the future search for the causes of PVC instability. The next paper by Světlý (29) explains a proposed chemical mechanism of "instability transfer", which is based on initiation due to the presence of the carbonyl group.

These general models are the only ones discussed in the literature so far. The following explanations should include these models in analysis, but before we discuss the kinetics of PVC degradation further, let us review some general trends of reactions which occur when PVC is thermally treated. Many authors have presented HCl emission versus degradation time. Abbås (30) shows the HCl emission process as a linear function of time (Fig. 2.7). Similar linear relationships are also presented in Wang's paper (31). In one recent paper (8), the HCl emission is not linear, as can be seen from Fig. 2.8.

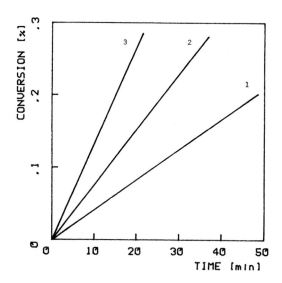

Fig.2.7. The degree of PVC dehydrochlorination as a function of heating time at different temperatures. 1 - 443K, 2 - 453K, 3 - 463K under N_2. (Modified from Ref. 30.)

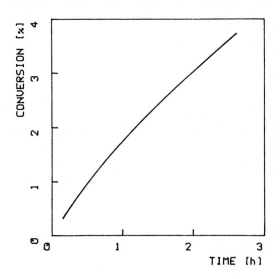

Fig.2.8. HCl loss as a function of time of thermal degradation at 463K under N_2. (Modified from Ref. 8.)

These two examples were shown together since some authors (28) believe that the rate of HCl emission should be linear, which would implicate a very simple mechanism of degradation, that is, in fact, related to many different factors. A much broader discussion of HCl emission can be found in Chapters Five and Six as it is related to the effect of the physical state barrier and methods of determination. For these explanations we should not simplify the HCl emission to a straight line, but try to find criteria for any course of changes.

The molecular weight of polymer increases with HCl emission, as can be seen from Fig 2.9 (32). This trend of gradual increase in molecular weight need not be permanent, as can be seen from the data presented by Varma (33) (Fig. 2.10).

It seems that both crosslinking and chain scissions are in a kind of an equilibrium that can change depending on conditions of degradation such as temperature, stage of degradation and HCl concentration. The number of polyenes having varying number of conjugated double bonds was studied by Abbås (19) (Fig. 2.11).

76

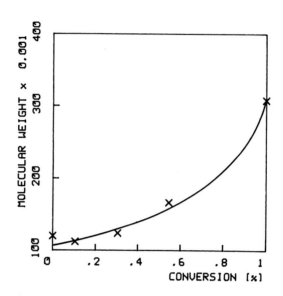

Fig.2.9. Weight-average molecular weight change as a function of conversion. Heating at 463K (N_2). (Modified from Ref. 32.)

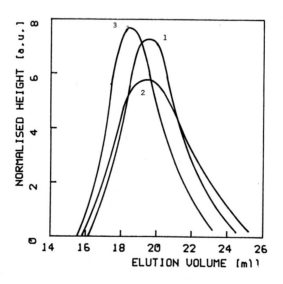

Fig.2.10. Molecular weight distribution of PVC degraded at 473K (N_2) at various time durations. 1 - 60 min, 2 - 120 min, 3 - 180 min. (Modified from Ref. 33.)

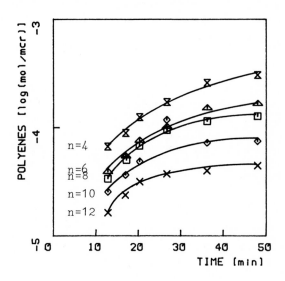

Fig.2.11. Log concentration of polyenes as a function of degradation time at 463K (N_2). (Modified from Ref. 19.)

These values were taken from UV measurement, and one should remember that they apply only to polymer in soluble form, while the fraction which already formed a gel might be missing. This could account for the decreasing number of polyenes with degradation time. We can consider such a probability based on data shown in Fig. 2.12 from the same work (19).

We can see that at the higher temperature of degradation, the average length of polyene is smaller; as calculated by Abbås (19), at 443K it is 14, while at 463K, it is 11, which should show a higher probability of polyene growth termination coinciding with temperature increase. Other studies corroborate that the number of polyenes does not decrease with time but rather increases (Fig. 2.13) (34). The radicals measured by ESR were identified as being of the polyenyl type, and we can see that their concentration increases exponentially. Surprisingly, the trends for long and short branches are just the opposite to each other (Figs. 2.14 and 2.15).

78

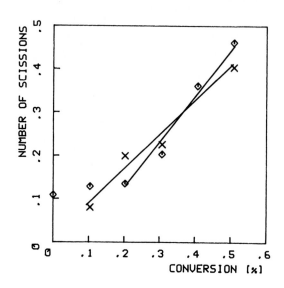

Fig.2.12. Number of scissions per 1000 mer units after ozonolysis versus conversion. ×- 443K, ◇ - 463K. (Modified from Ref. 19.)

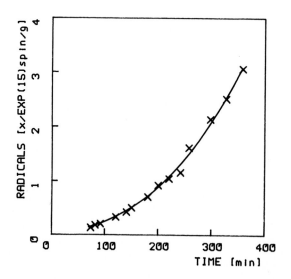

Fig.2.13. Average radical concentration versus heating time at 453K (N_2). (Modified from Ref. 34.)

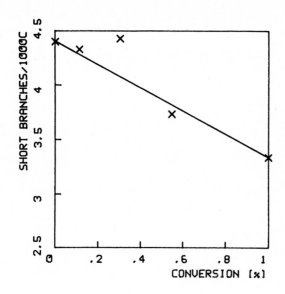

Fig.2.14. The number of short branches per 1000 carbon atoms versus conversion. Degradation temperature 463K (N_2). (Modified from Ref.32.)

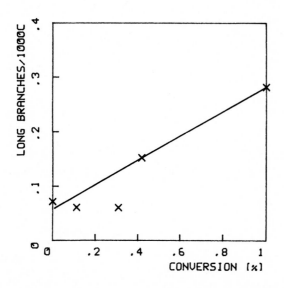

Fig.2.15. The number of long branches per 1000 carbon atoms versus conversion. Degradation temperature 463K (N_2). (Modified from Ref.32.)

2.2.1. Early stages at lower temperatures

Here we shall try to analyze the first few minutes of the degradation process, paying careful attention to initiation at temperatures at which pyrolytic changes do not occur, which would further complicate the mechanism. From the above discussion, we can single out the following models of kinetic change occurring during PVC thermal degradation:

1. First, molecules of HCl are eliminated, due to randomly distributed energy sufficient to cause conversion; then the process is catalyzed by the presence of HCl:

$$-CH_2-CHCl-CH_2-CHCl-CH_2- \xrightarrow{\Delta} -CH_2-CHCl-CH=CH-CH_2- + HCl \longrightarrow$$

$$[-CH_2-CHCl-CH_2-CHCl-CH_2-]\cdot HCl \longrightarrow -CH_2-CHCl-CH=CH-CH_2- + 2HCl$$

2. Initiation occurs due to HCl elimination by randomly distributed energy excess, and then the allylic structures provide centers for zip elimination:

$$-CH_2-CHCl-CH_2-CHCl-CH_2- \xrightarrow{\Delta} -CH=CH-CH_2-CHCl-CH_2- + HCl$$

$$\xrightarrow{\Delta} -CH=CH-CH=CH-CH_2- + 2 \ HCl$$

3. The chain includes labile structures (e.g., β-chloroallylic groups) which can initiate the process, and then the dehydrochlorination process occurs due to the propagation of structures with conjugated double bonds:

$$-CH=CH-CH_2-CHCl-CH_2-CHCl- \xrightarrow[-HCl]{\Delta} -CH=CH-CH=CH-CH_2-CHCl-$$

$$\xrightarrow[-HCl]{\Delta} (-CH=CH-)_3$$

4. The chain contains the labile structures (e.g., carbonyl-chloroallylic groups) which are not exhausted in a single initiation act but which can be transferred to another part of the chain to initiate new centers of zip

dehydrochlorination:

5. The radical mechanism of initiation proposed long ago by Winkler (35) and Arlman (36) also seems to be relevant. Three basic steps are involved: generation of radical, abstraction of hydrogen and production of polyene with simultaneous formation of chlorine radical:

$$R^{\cdot} + -CH_2-CHCl-CH_2-CHCl-CH_2- \longrightarrow RH + -CH_2-CHCl-\overset{\cdot}{C}H-CHCl-CH_2-$$

$$\longrightarrow -CH_2-CHCl-CH=CH-CH_2- + Cl^{\cdot}$$

When we analyze these models from the point of view of different polymer types, we can see that models one and two do not support a large differences in dehydrochlorination rate among various polymers, since elimination would be related to the type of bonds and energy supply.

If we analyze these models from the perspective of reaction kinetics, we can see that in the case of models one and two, the reaction rate grows exponentially since new centers are formed throughout the degradation process that will contribute to the increase of active centers of polyene growth. In the case of model three, reaction kinetics should proceed in the opposite direction. Initially many centers are formed from labile groups; then, when the polyenes are ready to terminate, the dehydrochlorination rate should drop down. Models four and five are quite similar so far as the dehydrochlorination rate is concerned. In both cases, there should be an accumulation of active centers; therefore, the rate of dehydrochlorination should grow steadily if labile structure transfer or abstraction of hydrogen is sufficiently effective. In model four there is no mechanism which can stop or decrease the rate of the transfer process while it exists in the case of the radical mechanism, since the only effective hydrogen abstraction is from the secondary carbon atom containing two hydrogens.

We can see, therefore, that the dehydrochlorination rate is either constant or slightly decreasing, and no one model alone can explain this experimental fact. Considering the concentration of polyenes and radicals (Figs. 2.12 and 2.13), we come to the same conclusion. Taking into account polyene length distribution (Fig. 2.11), we have to conclude that initiation is slower than propagation, since almost from the beginning we have polyenes of various length and their concentration seems to be parallel throughout the long period of degradation, which means that rates of initiation and termination are well balanced, at least at the beginning of degradation. The results presented in Fig. 2.11 should be regarded as good proof that the labile structures are really operative in the mechanism of PVC initiation. Světlý (28, 29) is correct that there must be another mechanism of initiation in the course of the further degradation process; otherwise the dehydrochlorination rate should drop. One valid argument, however, counters his proposal that carbonyl group transfer participation maintains the dehydrochlorination rate at a higher level. If this were true, how could we explain the mechanism of action of such stabilizers as phosphites or some organotins, which are able to reduce drastically the carbonyl group level, especially after the initial few minutes of degradation, even in the presence of oxygen? Does this mean that they are able to stop dehydrochlorination completely?

Doubtless Světlý´s hypothesis (28, 29) should be regarded as an important proposal, but not necessarily concerning carbonyl group participation. Perhaps the dehydrochlorination rate decrease due to polyene termination is increased by the catalytic effect of HCl and participation of the radical mechanism in activating new centers of dehydrochlorination. We cannot regard the mechanism of dehydrochlorination as homogeneous throughout the whole process of degradation. There are definitely steps in which some mechanisms are more important than others, and one such is initiation at the beginning of the process due to the presence of labile structures, especially the β-chloroallyl groups.

2.2.2. The effect of HCl

Only a few researchers (37, 38) fail to agree at the moment that HCl has a catalytic effect under inert conditions; otherwise all investigators working in this field seem convinced that HCl participates in dehydrochlorination catalysis in the presence of oxygen and in inert atmosphere. This unanimity of opinion should indicate that results are sufficient to support this belief. But this is not exactly the case because of the major difficulty of how to differentiate between HCl which is produced due to temperature and that produced due to autocatalysis. For this reason a researcher attempting to carry out measurements related to this phenomenon is forced to create experimental conditions which are artificial as compared with conditions in the normal degradation procedure. Frequently one must look for new techniques of measurement that otherwise are not common in PVC degradation studies.

The simplest technique of measuring the HCl evolved in an atmosphere of HCl, introduced from outside of the system, was adapted by Hjertberg (39) who simply controlled the sample weight being degraded for a specific time interval under nitrogen containing 0-40% HCl. Cooray (40) has done studies in solution which included a controlled concentration of dissolved HCl; polymer changes were not monitored, but the model peroxides were determined. Amer (41) has designed equipment consisting of the degradation vessel connected with a quartz tube inserted in the optical path of the IR spectrophotometer, which measured the concentration of HCl as a sum of that introduced and emitted.

The effect of HCl on sample weight loss (practically HCl) can be seen from Fig. 2.16. The effect of HCl on the dehydrochlorination rate looks very large, as does the concentration of HCl under which measurements were done. One must consider that the diffusion process is gradient controlled; therefore, the very high concentration of HCl in the surrounding atmosphere does not mean a proportional change in the concentration in the sample. Similar results by Amer (41) show that the higher additions of HCl do not adequately increase the dehydrochlorination rate, but there is a certain limit of concentration which stimulates the dehydrochlorination; then the dehydrochlorination rate drops down (25 cm Hg) (Fig. 2.17).

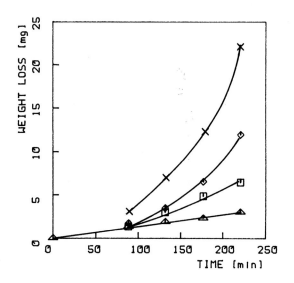

Fig.2.16. Sample weight loss versus degradation time at varying HCl concentrations. △ - control, □ - 10% HCl, ◇ - 20, × - 40. (Modified from Ref. 39.)

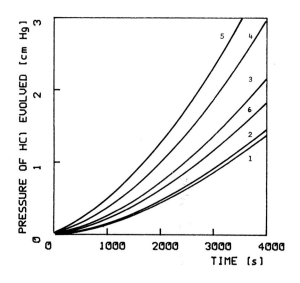

Fig.2.17. Pressure of HCl evolved versus degradation time at varying pressures of HCl added. Degradation temp. 463K (N_2). 1 - control, 2 - 1 cm Hg, 3 - 2, 4 - 5, 5 - 15, 6 - 25. (Modified from Ref. 41.)

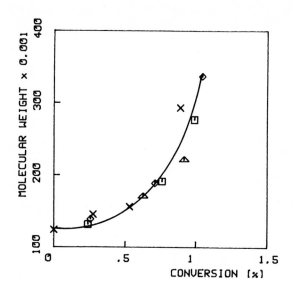

Fig.2.18. The relationship between molecular weight and conversion at various concentrations of HCl. × - control, Δ - 10% HCl, ◇ - 20%, □ - 40%. (Modified from Ref. 39.)

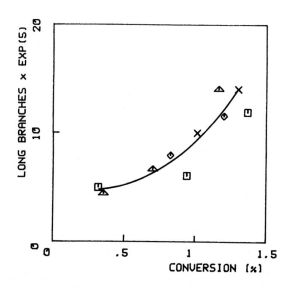

Fig.2.19. Concentration of long branches versus conversion. □ - control, ◇ - 10% HCl, × - 20%, Δ - 40%. (Modified from Ref. 39.)

Hjertberg´s work also gives data on the effect of the presence of HCl on the molecular weight of PVC and the number of long branches during degradation (Figs. 2.18 and 2.19). Regardless of HCl concentration, points form the same relationship, which means that there is no influence of HCl either on molecular weight change or long branch concentration, both of which are evidently related to conversion degree.

The polyene sequence length at different concentrations of HCl is given by Fig. 2.20.

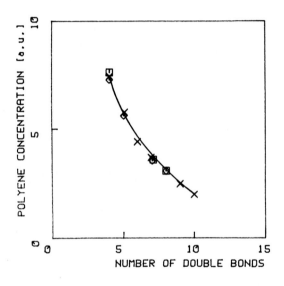

Fig.2.20. Normalized polyene distribution of samples degraded under varying concentrations of HCl. × - 10% HCl, ◇ - 20%, □ - 40%.(Modified from Ref. 39.)

The presence of HCl does not seem to affect polyene distribution, but at the same time Hjertberg found that with conversion the number of polyenes increases when the sample is degraded in nitrogen. If the presence of HCl does not affect polyene distribution, it probably affects initiation, since more HCl is produced. Unfortunately, no mechanism known at the moment can explain this process of initiation involving HCl catalysis. Hjertberg assumes that the HCl operates through the formation of a cyclic intermediate, as follows:

$$-CHCl-CH-CH-CH_2-CHCl- \longrightarrow -CHCl-CH=CH-CH_2-CHCl- + 2HCl$$

This mechanism and process of HCl catalysis might be useful in explaining the dehydrochlorination rate without the necessity of applying Světlý's model (28, 29) of transportation of labile structure during the course of degradation.

The effect of HCl is even more complicated when one considers degradation in the presence of oxygen. One would think that the increase in HCl concentration should cause a proportional increase in PVC instability. From Cooray's data (40), it is postulated that HCl decreases the concentration of peroxides (model substance in solution), which is believed to be a cause of catalytic free-radical redox decomposition. This process probably leads to the enhanced initiation of dehydrochlorination by the radicals formed.

Summarizing the above discussion, three points should be stressed:

1. that available results implicate the participation of HCl in dehydrochlorination autocatalysis,
2. the effect of HCl is on initiation rather than on propagation or termination rates,
3. results are still tentative and require confirmation by more experimental data.

2.2.3. Conjugated double-bonds formation

Based on the available literature, we discussed the possible reasons for initiation at the beginning of thermal degradation reactions and during the course of further dehydrochlorination when new centers are established that are to be propagated to form longer sequences of conjugated double bonds. We could see (Fig. 2.11) that polyene concentrations of varying length are parallel throughout the process of degradation. Fig. 2.12 shows that, all logic to the contrary, shorter polyenes are formed when temperature is increased. Finally, Fig. 2.20 tells us that the concentration of HCl does not seem to have an effect on polyene length distribution. These observations seem to show that, first of all, there must be a mechanism of polyene termination that does not allow them to grow above a certain limit. Many authors have tried to define this limiting value; for example, Braun (42) said that conjugated double bonds are in the range of 5 to 10; Kelen (12), 6 to 11; Palma (43), 4 to 12; Daniels (44), 6 to 12; Owen (45), 11 to 13; and Atchison (46), 4 to 16. Regardless of the small differences, the result of each investigator performing his experiment under different conditions, one conclusion is always the same: polyenes are terminated quite early. Interestingly, if we look at naturally-occurring polyenes, we can see that their most frequent length is eleven conjugated double bonds, and none has more than fifteen. It looks as if Nature, which is always very economical in its goals, also does not try to exceed a certain limit, which happens to be similar to that which we observe in PVC degradation. Cases have been reported of longer polyenes; one (referred to in the next chapter) by Shindo (47), presented results from chemical dehydrochlorination of PVC (Fig. 3.25). Numerous results have been reported in the case of polyacetylenes, but these are mainly based on guesswork, since it is hard to define polyene length exactly in a material which is built up only from polyenes divided by some chain irregularities. The literature on polyacetylenes yields the following figures: Harada (48) claims that polyacetylene is composed of finite polyene sequences with 4 to 400 double bonds; an average conjugation of 30 double bonds was determined by Kuzmany (49). Rives-Arnau (50) polymerized samples with up to 20 conjugated double bonds in sequence, while Wegner (51) found polyacetylene to be composed of rather short polyenes.

As a matter of fact, neither chemical dehydrochlorination studies nor data from acetylene polymerization are particularly relevant for us, since both mechanisms of polyene formation are entirely different from those with which we are dealing. On the other hand, we should be more concerned with the reason for termination rather than with a quantitative approach to polyene formation. The other striking feature is that polyenes are shorter when temperature is increased. Two questions must, therefore, be addressed:

1. Why are the polyenes terminated?
2. Why does temperature increase cause the polyenes to be shorter?

There could be no further discussion if not for the pioneering work of Haddon and Starnes (52, 53), who applied molecular orbital calculations in order to explain this phenomenon. One can hardly imagine experimental techniques, at the moment, which could possibly answer these two important questions. These investigators applied MINDO/3 and ab initio STO-3G calculations to analyze the electronic structures, equilibrium geometries and energy of the ground states of polyenyl cations and neutral polyenes. We will omit some technical aspects of their papers (52, 53), since these can be found in the originals, and concentrate instead on the results of this important work.

All bond angles and C-H bond length were assumed from available data, but bond length between any two carbon atoms, although initially imputed, was approximated by parabolic interpolation. Studies were done for linear polyenes of structure:

with two to ten carbon atoms, and for polyenyl cations:

with three to 11 carbon atoms.

The conjugation energies of linear polyenes are given in Fig. 2.21. The energy pattern shows that conjugation is favored, and that the energy depends on the number of conjugated double bonds. The conjugation energy of linear polyenyl cation is given by Fig. 2.22. Since the reaction proceeds with a loss of conjugation and charge dispersal, the energy is higher and positive.

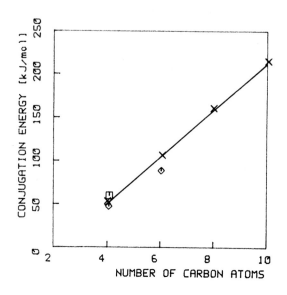

Fig.2.21. Energy of conjugation for linear polyenes. ×- STO-3G, ◇ - 4-31G, □ - experimental. (Modified from Ref. 53.)

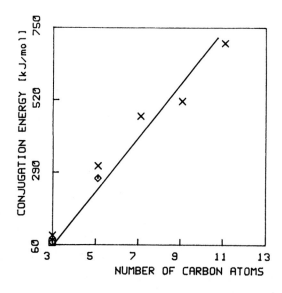

Fig.2.22. Energy of conjugation for linear polyenyl cations. × - STO-3G, ◇ - 4-31G, □ - experimental. (Modified from Ref. 53.)

An alteration of charge density was observed; this means that even-numbered carbon atoms preferred to have a different sign of charge density than their odd neighbors.

The authors (52, 53) conclude that the forward reaction step is favored by unsaturation:

Since the charge is better established in the polyene center, there is the possibility of migration of an ion pair:

With increasing stabilization of the ion pair by unsaturation, the migration of the ion pair towards the center of the chain becomes more difficult, as does the abstraction of the proton, since the chloride ion is no longer in the neighborhood of the methylene group. Haddon and Starnes (52, 53) noted that the charge density on methylene protons decreases along with increased unsaturation, which also does not favor elimination. This seems to explain why polyenes do not grow beyond a certain point.

Attracted by the success of these studies, we have decided to repeat the calculations using throughout the MINDO/3 method which allows us to compare longer polyene cations and obtain comparable data for all polyene lengths (54). Our calculations include polyenyl cations from C_3 to C_{17} for the following structure:

As the structure is symmetrical, the data for segment 9-1 are equivalent to those for segment 1-17, and only segment 1-9 will be given in the data below. Table 2.5 shows the length of carbon-carbon bonds, while Table 2.6 shows the total bond order.

TABLE 2.5.

The length of carbon-carbon bonds, Å. (54)

Cation	Bond length							
	9-8	8-7	7-6	6-5	5-4	4-3	3-2	1-2
C_3								1.402
C_5							1.381	1.432
C_7						1.370	1.440	1.411
C_9					1.376	1.445	1.396	1.420
C_{11}				1.364	1.455	1.389	1.434	1.411
C_{13}			1.360	1.460	1.382	1.442	1.399	1.420
C_{15}		1.359	1.463	1.377	1.448	1.392	1.430	1.411
C_{17}	1.359	1.463	1.375	1.451	1.386	1.438	1.401	1.419

TABLE 2.6.

The total bond order for polyene cations of varying length (54).

Cation	Bond order (total)							
	9-8	8-7	7-6	6-5	5-4	4-3	3-2	2-1
C_3								1.703
C_5							1.813	1.541
C_7						1.864	1.460	1.666
C_9					1.890	1.414	1.736	1.580
C_{11}				1.913	1.371	1.788	1.513	1.657
C_{13}			1.927	1.343	1.824	1.461	1.717	1.590
C_{15}		1.937	1.323	1.805	1.420	1.763	1.533	1.653
C_{17}	1.943	1.311	1.868	1.390	1.799	1.484	1.708	1.596

We can see an interesting phenomenon for polyenic cations of an even and odd number of conjugated double bonds. If we were to assign the formal double and single bonds to the results, the structures would appear as follows:

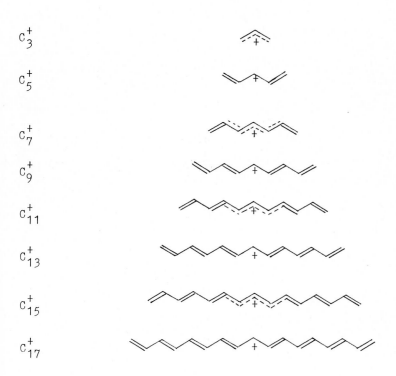

C_3^+

C_5^+

C_7^+

C_9^+

C_{11}^+

C_{13}^+

C_{15}^+

C_{17}^+

The longer the polyene, the more the double-bond character is observed, which may be related to the structure of polyenic cation assumed in our calculations, but Yamabe's work (55) shows that this is not the case. Yamabe has made similar calculations for C_{12} polyene and its single and double cations, as follows:

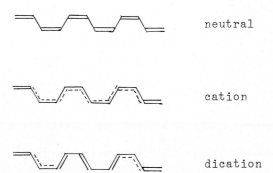

neutral

cation

dication

The bond order (total) is as follows:

Bond:	1-2	2-3	3-4	4-5	5-6	6-7
neutral	1.96	1.28	1.92	1.30	1.91	1.30
cation	1.85	1.42	1.69	1.56	1.62	1.62
dication	1.80	1.56	1.51	1.80	1.34	1.88

The electron densities on particular carbon atoms from Stȩpień's and Wypych's work (54) are given in Table 2.7.

TABLE 2.7.

The electron density on each carbon atom (54).

Cation									
	9	8	7	6	5	4	3	2	1
C_3								3.640	4.259
C_5							3.762	4.225	3.696
C_7						3.829	4.208	3.753	4.213
C_9					3.870	4.194	3.799	4.207	3.773
C_{11}				3.907	4.179	3.836	4.198	3.792	4.211
C_{13}			3.932	4.167	3.868	4.187	3.817	4.208	3.797
C_{15}		3.952	4.157	3.895	4.175	3.843	4.201	3.809	4.211
C_{17}	3.967	4.148	3.917	4.163	3.867	4.193	3.826	4.209	3.809

The electron density follows the same pattern as the data given above. It is also evident that when polyene grows, the charge density differences between neighboring carbon atoms become smaller. The same pattern is favored when these differences between atoms in the center and close to the terminal carbons are compared.

The bond length and electron density for terminal carbons are given in Figs. 2.23 and 2.24, respectively.

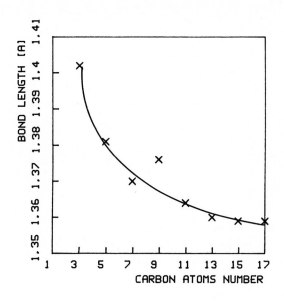

Fig.2.23. Terminal double bond length in A versus polyenic cation length (54).

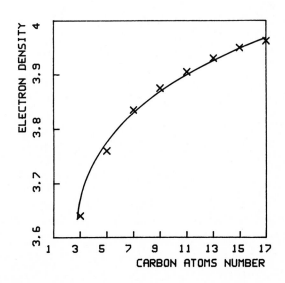

Fig.2.24. Electron density on terminal carbon versus polyenic cation length (54).

The above seems to explain why polyenes have to terminate. As double bond length is already stabilized, and, more importantly, electron density is approaching value 4, there is gradually less of an effect on the neighboring monomer unit, which becomes a regular unit in the PVC chain, and thereby the HCl can be eliminated (as from any other regular unit) only by random elimination.

In addition, we have calculated the ionization potential given in Fig. 2.25.

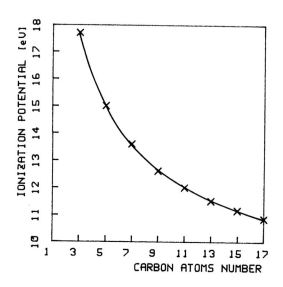

Fig.2.25. Ionization potential versus polyenic cation length (54).

Since the ionization potential value decreases with polyene increase, it may explain why temperature increase and the presence of oxygen are so fatal for longer polyenes. It is also interesting to notice that Bernier (56) has suggested that crosslinks appear in polyacetylene during thermal treatment as a consequence of the trans sequence presence having an odd number of carbon atoms with unpaired electrons. Perhaps this explains why crosslinks are formed.

Stępień's and Wypych's explanations (54), although slightly different from those presented by Haddon and Starnes (53), lead to the same conclusion: that there is a natural mechanism which does not allow polyenes to grow above a certain length.

2.2.4. The effect of oxygen in the thermal degradation process

The following discussion includes an explanation of thermooxidative changes by comparison with similar changes in inert atmosphere in order to clarify the interrelationship of both processes.

The oxygen intake of samples exposed to oxygen at atmospheric pressure during thermal degradation is given by Fig. 2.26.

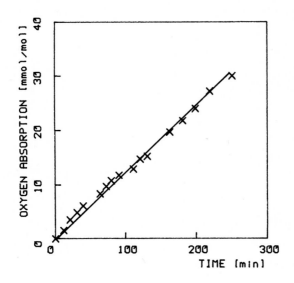

Fig.2.26. Oxygen absorbed versus degradation time at 453K. (Modified from Ref 57.)

The oxygen absorbed increases the concentration of peroxides as shown by Fig 2.27. Clearly, temperature has a detrimental effect on this process, and the process has three phases. After initial growth, peroxide concentration drops to zero only to grow again more rapidly in the course of further thermooxidation. Oxygen intake also accelerates spin generation increase when compared to a sample degraded in nitrogen (Fig. 2.28). Interestingly, a certain induction period is needed before spin generation can be effectively monitored. Actually, however, the course of changes presented in Fig. 2.27 seems illogical. Cooray´s work (40), fortunately, seems to present an answer. (Figs. 2.29 and 2.30.)

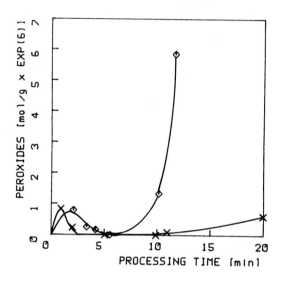

Fig.2.27. The effect of temperature on formation of peroxides during thermooxidation. × – 443K, ◇ – 483K. (Modified from Ref. 58.)

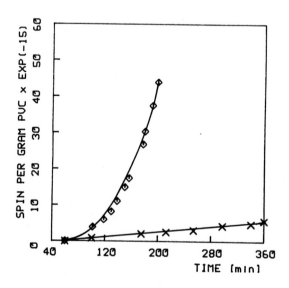

Fig.2.28. The effect of degradation atmosphere on spin generation during thermal degradation at 467K. × – in nitrogen, ◇ – in oxygen. (Modified from Ref. 59.)

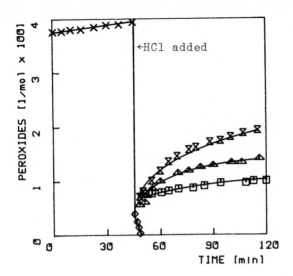

Fig.2.29. The effect of HCl addition on cumene hydroperoxide concentration at various HCl concentrations (M x 100). ◇ - 0.375, □ - 3.75, Δ - 7, X - 15. (Modified from Ref. 40.)

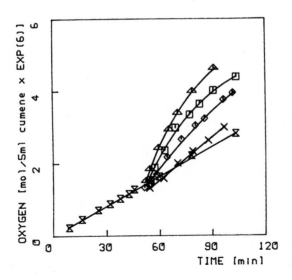

Fig.2.30. The effect of HCl addition on oxygen absorbed, depending on cumene hydroperoxide concentration (M x 100). x - 0.375, ◇ - 3.75, □ - 15, Δ - 30. (Modified from Ref. 40.)

Introduction of HCl in a concentration lower than that of cumene hydroperoxide dissolved in chlorobenzene caused almost immediate decomposition of hydroperoxide. Along with an increase in HCl concentration, the hydroperoxide concentration, after an initial rapid decrease, grew faster. The oxygen absorption rate was higher when the concentration of HCl was increased. The peroxide decomposition is explained by reaction:

$$
\underset{\substack{|\\ \text{H-Cl}}}{R-\overset{|}{\underset{|}{C}}-O-O-H} \longrightarrow R-\overset{|}{\underset{|}{C}}-O^{\cdot} + H_2O + {}^{\cdot}Cl
$$

which shows that radicals are formed. The increased concentration of HCl stimulates the formation of peroxides, which, in turn, produces more free radicals, and thus the degradation process is enhanced. Fig. 2.5 above compares the dehydrochlorination rate in inert atmosphere and when the sample was preoxidized. Doubtless, oxygen intake increases the dehydrochlorination rate. We have an almost ready answer for how this rate is increased since we have established that radicals are formed from peroxides. These radicals may certainly act in the formation of a new center of initiation by abstracting hydrogen from regular units of the PVC chain, but how are they actually formed?

Tüdős (57), studying polyene consumption during oxidation time, found that the following relationship is fulfilled (Fig. 2.31). Thus peroxyradicals are ready to react with any double bond with the same probability. Gupta (59) performed an interesting experiment, and his data are given by Fig. 2.32. Thus even at room temperature, oxygen can produce radicals in a reversible way, which, in turn, shows that radicals can be produced from polyenes, as predicted by Druesdow (60) long ago:

$$
-CH=CH-CH=CH- + O_2 \longrightarrow \underset{\substack{|\quad\quad\quad|\\ O\text{------}O}}{-CH-CH=CH-CH-} \longrightarrow \underset{\substack{|\quad\quad\quad|\\ O^{\cdot}\quad\quad O^{\cdot}}}{-CH-CH=CH-CH-}
$$

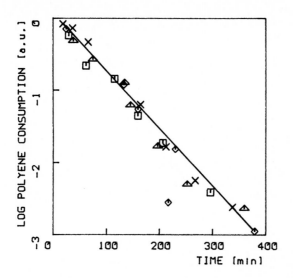

Fig.2.31. Logarithmic plot of polyene consumption of varying length versus oxidation time. ×- 5, ◇ - 7, □ - 9, Δ - 11. (Modified from Ref. 57.)

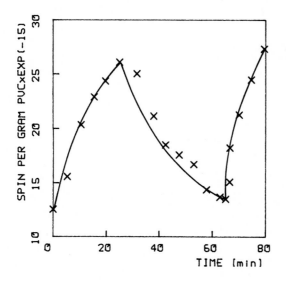

Fig.2.32. The effect of oxygen at room temperature on spin previously generated by heating PVC in N_2 at 523K. At point time 0, oxygen was introduced, replaced by nitrogen after about 30 min, which was again replaced by oxygen at about 65 min of process duration. (Modified from Ref. 59.)

Druesdow´s mechanism and the above-mentioned radical generation
process sufficiently explain why the dehydrochlorination rate
increases when the sample is degraded in the presence of oxygen.

Gupta (59) determined gel formation in PVC degraded in inert
atmosphere and in the presence of oxygen (Fig. 2.33).

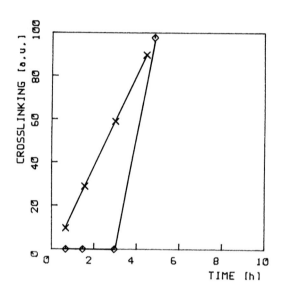

Fig.2.33. Crosslinking versus heating at temperature 463K. ◇ -
oxygen, × - nitrogen. (Modified from Ref. 59.)

Apparently the crosslinking process is not favored for a long
period of heating in the presence of oxygen. Only at very
advanced stages of degradation may one observe sudden
crosslinking. Many researchers (30, 36, 60-62) have found the
opposite trend, i.e., chain scission almost from the beginning of
the thermal oxidation process.

The above discussion shows explicitly that degradation in inert
atmosphere and in the presence of oxygen differs in several
points. First, we have the presence of free radicals, which can
cause additional initiation of dehydrochlorination centers.
Because polyenes might be oxidized, we have no parity between the
degree of dehydrochlorination and sample color, as more or less
exists in the case of degradation in inert atmosphere. Abbas
(30) believes that some growing polyenes can be deactivated by

the following reaction with oxygen:

$$-CH=CH-CHCl-CH_2- \xrightarrow{O_2} -CH=CH-CCl-CH_2- \longrightarrow$$
$$\underset{OOH}{|}$$

$$-CH=CH-CCl-CH_2- + \ ^{\bullet}OH \longrightarrow -CH=CH-CCl + \ ^{\bullet}CH_2-$$
$$\underset{O^{\bullet}}{|} \qquad\qquad\qquad \underset{O}{||}$$

which cannot much decrease the dehydrochlorination rate since new radicals are formed in the process that may cause further initiations.

The most severe changes occur in mechanical properties since the molecular weight of polymer decreases, which is precisely the opposite trend to degradation in inert atmosphere wherein chain scissions have rather marginal importance and crosslinking is the predominant factor. Similar conclusion can be reached from work on polyacetylenes (63, 64). Here electrical conductivity measurement revealed that oxidation of polymer destroys conjugation and reduces chain length, changes which finally make the polymer more brittle.

Although we have been able to find many explanations for observations made in the course of thermal oxidation which could be smoothly put in a logical sequence, this does not mean that the process is so well understood. We can easily conclude that what is really quite well studied is the possibility of initiation of new active centers of degradation, which explains why the thermooxidative process is faster than dehydrochlorination in inert atmosphere. On the other hand, if we were able to construct models of all these mechanisms of initiation and express them in numerical form, we would soon discover that PVC should disintegrate almost immediately in the presence of oxygen since so many factors are against its stability. This shows exactly what is missing in our studies: knowledge of propagation and termination mechanisms and kinetics. The absence of this knowledge justifies certain reservations concerning the validity of current explanations. This situation is not surprising if we consider that many questions remain with regard to simple dehydrochlorination in inert atmosphere; why then should everything be so clear if the model is further complicated by an additional set of parameters?

2.2.5. Volatile products of the thermal degradation process

Kelen (12, 65) reported the formation of benzene during PVC degradation in inert atmosphere at 453K. His results are given in Fig. 2.34.

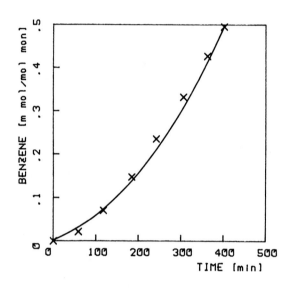

Fig.2.34. The number of benzene molecules evolved per monomer unit during thermal degradation. (Modified from Ref. 65.)

When one compares the amount of benzene formed with the HCl molecules evolved based on the figure from Kelen's other paper (12), it looks as if his results suggest that out of 15 double bonds formed at a degradation temperature of 453K, one is used for benzene formation. Thus, benzene formation would be a major process in PVC thermal degradation. From the linearity of the relationship, we should also consider this process as being part of degradation; in other words, there should be a specific mechanism parallel to dehydrochlorination which causes benzene formation from the beginning of the process, when particular structures are available in polymer chain. Such mechanisms have been proposed by Starnes (66), as explained by the following chemical equations:

where:

R_1, R_2 - -CH_2-, -CHCl- or -CH=CH-

which means that either an intermolecular or intramolecular
Diels-Adler mechanism is suggested. A similar mechanism is
proposed by Kuzmany (67) for polyacetylenes, while Ivan (8)
proposes addition involving prior cyclopentadienylation.

Based on these explanations, one might feel that the problem is
solved, but such is not the case. First, detailed studies by
Liebman (68) revealed that the optimal temperature of benzene
production equals 598K and initial detection of benzene is at
523K. Ballistreri (69) has also detected benzene formation only
above 473K. Using the most sophisticated equipment for this type
of study, both investigators could not have missed benzene if it
had been produced as effectively as Kelen claimed. The other
reservation comes from the mechanism of the Diels-Adler reaction,
which requires that some specific conditions for diene and
dienophile be fulfilled (compare Chapter One). Moreover, the fact
is related to polymer conformation, as it is not clear how
cis-transoid and trans-cisoid conformations can be achieved
simultaneously in two contacting chains if the polyene structure
is expected to be all-trans, as explained in Chapter One. If
benzene is formed at this stage of degradation (below 473K), it
should really be part of the dehydrochlorination mechanism, since
only when a double bond is just formed (before isomerization
occurs) can such conformation be present. The other explanation
is that branches naturally occurring in the chain are involved in
the process according to a presently unknown mechanism. Indeed,
it was shown earlier that the number of short branches is
decreased during the degradation process, which eventually can be
connected with benzene formation. Clearly, further studies are
needed in order to resolve these inconsistences.

2.2.6. The mechanism of thermal degradation below 473K

Throughout Chapters One and Two we have discussed the observations and doubts related to particular elements which together form the sequence of events that could be called the mechanism of PVC thermal degradation. Now we must summarize the previous paragraphs. It is customary to discuss separately the mechanism of PVC thermal degradation under inert atmosphere and in the presence of oxygen, but if we take into consideration that practically every processing method is realized in the presence of a certain concentration of oxygen, we should view the results obtained under inert gases as model conditions which help us to better understand the more complicated and sometimes different nature of degradation which PVC undergoes when subjected to practical conditions.

It is also customary here to argue whether the mechanism of PVC degradation should be viewed as a radical or an ionic one, but we do not support this type of discussion, especially when an overall look at the mechanism is concerned, since the answer is clearly that they are both in operation. There is sufficient evidence to suggest that if HCl is present, it autocatalyzes PVC thermal degradation, at which point an ionic mechanism goes into operation. On the other hand, so many proofs for radical participation were cited, products identified, kinetics measured, that one can only exclude a radical mechanism out of stubborness.

Initiation

Evidence from kinetic studies of the process of PVC thermal degradation shows that one has to distinguish between two types of initiation, one at the beginning of heating and another throughout the period of degradation. The first initiation process is well supported by experimental data and is most likely due to the presence of internal unsaturations leading to both α - and β -allylic chlorine. Because there is evidence from ESR studies that radicals are present from the beginning of the heating process, there is a very low electron density on both terminal carbons in the vicinity of the double bond. The polyene cation is a symmetric moiety, which means that there is practically no reason to discriminate between further zipping in either direction, since what the double bond does is to polarize the C-Cl bond to allow for easier elimination. Obviously, if better evidence were found confirming that carbonyl structures are really present next to the double bond, we should agree on their participation in the initiation mechanism since there is no reason to disqualify them. Other groups, such as tertiary carbon-bound chlorine, syndiotactic sequences and so on, proposed as possible initiation centers, should still be considered tentative, because the intermediate structures are unknown and the experimental data is not strong enough to determine their real effect.

Considering that early polyene termination is a proven fact, we have to agree that there must also be another mechanism of initiation operative throughout the degradation process that is able to replace active centers in order to keep degradation going. Random elimination from typical units of the PVC chain alone cannot be a useful explanation. Our choice is, therefore, limited either to migration of an active group able to initiate dehydrochlorination in a "healthy" part of the chain or to agreement that once HCl is present, it autocatalyzes the initiation. The first hypothesis has not yet been proved. Yet, how could we eventually explain HCl autocatalysis, if it does not affect polyene distribution? Therefore, if it cannot participate in either propagation or termination of polyenes, it has to be included in the operative mechanism of initiation throughout the degradation process.

Based on the higher probability that HCl autocatalysis may be the reason for continuing initiation, we assume that this effect is operative, but this does not exclude the possibility that migration of any group may also take place in initiation of thermal degradation. As far as the presence of oxygen is concerned, sufficient evidence exists to believe that the higher rate of PVC thermooxidation is explained by participation of peroxides and products of their almost immediate decomposition in more effective initiation of thermal dehydrochlorination. In the case of thermooxidative degradation, certainly more weight should be given to the radical process which was so explicitly demonstrated in Fig. 2.33.

Propagation

This step of the mechanism is naturally supported by charge distribution in the polyene structure, as demonstrated by molecular orbital calculations. It should not be regarded as a sequence of the same steps, since every additional step seems more difficult because the polyene center is a point of symmetry so far as consecutive properties of bonds are concerned. If, apart from the doubts expressed in paragraph 2.5, we assume the formation of benzene in the type of degradation under discussion, we suggest that the intramolecular cyclization is a part of the propagation mechanism due to this process´ structural constraints. There is no particular reference in the literature concerning the effect of oxygen´s presence on polyenes propagation, and this point is not clear. But it is hard to foresee any reasonable way in which oxygen would be able to affect the propagation step.

Termination

Here, again, we should see several reasons for polyene termination. Apart from the mechanism proposed in paragraph 2.3, due to which polyene growth ceases because of balanced charge density at terminal carbons when polyene exceeds a certain length, there is a mechanism which terminates the polyene due to crosslinking. It is also probable that the zipping reaction can be terminated at a branching point, but this is still subject to further investigation. Polyene growth is also stopped by chain scission due to thermooxidative reaction. Finally, some radical chain reactions leading to polyene propagation would be inhibited by the presence of oxygen due to transfer of excitation to oxygen. The temperature increase favors termination, most probably due to the lower ionization potential when polyene length is increased. In the presence of oxygen this termination would probably lead to chain scission; in its absence, to a crosslinking reaction.

Polyene degradation

Photobleaching studies have contributed to an understanding of the fate of polyene. It is well demonstrated that oxygen is ready to react with any double bond, no matter how long the conjugation. This shows that the thermal degradation process is essential to further stability of PVC material, since all weak structures are subjected to further conversion due to the presence of oxygen or to the photolytic processes.

2.3. PVC thermal degradation at high temperatures

PVC thermal degradation at high temperatures comprises two processes: pyrolysis and combustion. These processes are investigated for several reasons. First, the pyrolysis of polymers under controlled condition has been used for a long time as a useful analytical tool for their identification and structural analysis. Pyrolytic processes are studied because of the always-existing danger of combustion in the case of accidental fires; therefore, it is important to know at least the composition of the gases evolved, bearing in mind their possible toxicity. Waste materials, including polymers, are frequently decomposed by combustion, since biological methods proved inadequate. The problem of environmental pollution, therefore, plays an essential role in shaping these studies. Finally, some polymeric materials are hot-knife cut, which may cause emission of toxic fumes dangerous for workers in such environments.

The number of compounds which can be emitted during combustion seems to be unlimited, since one author found 75 products (70), and another, 59 compounds (71). There are four inorganic compounds in the case of combustion, i.e., HCl, H_2O, CO and CO_2, and just one in pyrolysis. Much discussion exists as to whether phosgene can be formed during pyrolysis or combustion. Brown (72) has done studies using four decomposition methods. The highest concentrations of phosgene (0.2-1.6 mg/g PVC) were generated in electric arc decomposition, while other methods produced 0.1-0.2 mg/g PVC.

Remaining volatile material comprises organic, mainly aromatic compounds, and what is important is that there are no oxygen-containing products in either combustion or pyrolysis, and hence degradation is a non-oxidative process (70, 71). Most chlorine is emitted in the form of HCl, and only very small quantities of vinyl chloride, being the only organic material containing chlorine, were detected. The aliphatic compounds comprise mainly methane, ethylene, propylene, propane, 1-butene, isopentane, 1-pentene, cyclopentene, cyclopentane, 1-hexene, hexane, all in very small quantities. Similar compounds are produced during polyacetylene pyrolysis (73).

A wide variety of aromatic compounds has been detected on pyrolysis of both polyacetylene and PVC. The main components include: benzene, toluene, xylene, styrene, indene, naphthalene, bisphenyl, fluorene and even anthracene. Both qualitative and quantitative composition of products is similar to the decomposition of polyacetylene and PVC, which clearly shows that there are at least two stages in the decomposition process. The first comprises HCl elimination; the second, formation of organic compounds. Actually, both processes overlap (69). There are two maxima of formation of organic compounds: one at about 560K, and the other at 620K. In the first maximum of emission, benzene is formed very efficiently. With temperature increase, the concentration of benzene in volatile products decreases, and others like naphthalene, anthracene and methylnaphthalene are produced more efficiently around the second maximum.

Lattimer (74) proposed mechanisms of formation of several aromatic compounds. Let us review two as examples:

Benzene

One can see that the mechanism proposed by Lattimer is the same as Starnes (67) proposed for intramolecular benzene formation during thermal degradation at lower temperatures.

Naphthalene

Generally, benzene, naphthalene, and probably also biphenyl and anthracene are thought to be formed exclusively via intramolecular cyclization reactions. Styrene and indene can be formed by the same mechanism or via transfer of the hydrogen atom between the chains. Intermolecular formation has been proposed for toluene and naphthalene. Aliphatic pyrolyzates are formed by chain scission.

Pyrolytic changes are closely related to the temperature program at which decomposition occurs. Also, the sample residence time related to its size is important.

REFERENCES

1. G. Kimura, K. Yamamoto, K. Sueyoshi and Y. Gosei, **Kagaku Kyokaishi**, 23(1965)136.
2. Z. Mayer, B. Obereigner and D. Lim, **J. Polym. Sci.**, C33(1971)289.
3. A.H. Frye and R.W. Horst, **J. Polym. Sci.**, 40(1959)419.
4. A.H. Frye and R.W. Horst, **J. Polym. Sci.**, 45(1960)1.
5. A.H. Frye, R.W. Horst and M.A. Paliobagis, **J. Polym. Sci.**, A2(1964)1765.
6. W.I. Bengough and M Onozuka, **Polymer**, 6(1965)625.
7. E.C. Buruianǎ, A. Airinei, G. Robilǎ and A. Caraculacu, **Polym. Bull.**, 3(1980)267.
8. B. Iván, J.P. Kennedy, T. Kelen and F. Tüdös, **J. Macromol. Sci.-Chem.**, A17(1982)1033.
9. A. Guyot, M. Bert, P. Burille, M.-F. Llauro and A. Michel, **Pure Appl. Chem.**, 53(1981)401.
10. T. Hjertberg and E.M. Sörvik, **J. Macromol. Sci.-Chem.**,

112

A17(1982)983.

11. D. Braun, **Pure Appl. Chem.**, 53(1981)549.

12. T. Kelen, **J. Macromol. Sci.-Chem.**, A12(1978)349.

13. K.S. Minsker, A.A. Berlin, V.V. Lisitsky and S.V. Kolesov, **Vyssokomol. Soed.**, A19(1977)32.

14. K.S. Minsker, V.V. Lisitsky, M. Kolinsky, Z. Vymazal, D.H. Bort, V.P. Lebedev, A.I. Petzena and E.M. Ilkayeva, **Plast. Massy**, 9(1977)44.

15. K.S. Minsker, V.V. Lisitsky, R.F. Gataulin and G.E. Zaikov, **Vyssokomol. Soed.**, B20(1978)571.

16. K.S. Minsker, V.V. Lisitsky, S.V. Kolesov and G.E. Zaikov, **J. Macromol. Sci.-Rev. Macromol. Chem.**, C20(1981)243.

17. K.S. Minsker, V.V. Lisitsky and G.E. Zaikov, **Vyssokomol. Soed.**, A23(1981)483.

18. Z. Mayer, **J. Macromol. Sci.-Rev. Macromol. Chem.**, C10(1974)263.

19. K.B. Abbås, **J. Macromol. Sci.-Chem.**, 12(1978)479.

20. K.S. Minsker, V.V. Lisitsky and G.E. Zaikov, **J. Vinyl Technol.**, 2(1980)77.

21. A. Guyot and A. Michel, **Dev. Polym. Stab.**, 2(1980)89.

22. M. Onozuka and M. Asahina, **Rev. Macromol. Chem.**, 4(1970)235.

23. D. Braun and M. Wolf, **Angew. Makromol. Chem.**, 70(1978)71.

24. A.A. Caraculacu, **Pure Appl. Chem.**, 53(1981)385.

25. S. Crawley and I.C. McNeill, **J. Polym. Sci., Polym. Chem. Ed.**, 26(1978)2593.

26. P. Šimon and L. Valko, **Coll. Czech. Chem. Commun.**, 47(1982)2336.

27. J.D. Danforth, J. Spiegel and J. Bloom, **J. Macromol. Sci.-Chem.**, A17(1982)1107.

28. J. Světlý, R. Lukáš and M. Kolinský, **Makromol. Chem.**, 180(1979)1363.

29. J. Světlý, R. Lukáš, J. Michalcová and M. Kolinský, **Makromol. Chem., Rapid Commun.**, 1(1980)247.

30. K.B. Abbås and E.M. Sörvik, **J. Appl. Polym. Sci.**, 17(1973)3567.

31. J.H. Wang and W.C. Tsai, **J. Chinese Inst. Chem. Eng.**, 10(1979)97.

32. K.B. Abbås and E.M. Sörvik, **J. Appl. Polym. Sci.**, 19(1975)2991.

33.   I.K.   Varma   and   K.K.   Sharma,   **Angew.   Makromol.   Chem.**,
      79(1979)147.

34.   J. Menczel, J. Varga, K. Juhass and M. Binett, **Period.**
      Polytech., **Chem. Eng.**, 22(1978)289.

35.   D.E. Winkler, **J. Polym. Sci.**, 35(1959)3.

36.   E.J. Arlman, **J. Polym. Sci.**, 12(1954)543.

37.   T. Kelen, G. Balint, G. Galambos and F. Tüdös, **J. Polym.**
      Sci., C33(1971)211.

38.   P. Bataille and B.T. Van, **J. Polym. Sci.**, **A-1**, 10(1972)1097.

39.   H. Hjertberg and E.M. Sörvik, **J. Appl. Polym. Sci.**,
      22(1978)2415.

40.   B.B. Cooray and G. Scott, **Eur. Polym. J.**, 16(1980)169.

41.   A.R. Amer and J.S. Shapiro, **J. Macromol. Sci.-Chem.**,
      A14(1980)185.

42.   D. Braun and M. Thallmeier, **Makromol. Chem.**, 99(1966)59.

43.   G. Palma and M. Carenza, **J. Appl. Polym. Sci.**, 16(1972)2485.

44.   V.D. Daniels and N.H. Rees, **J. Polym. Sci.**, **A-1**,
      12(1974)2115.

45.   E.D. Owen and I. Pasha, **J. Appl. Polym. Sci.**, 25(1980)2417.

46.   G.J. Atchison, **J. Appl. Polym. Sci.**, 7(1963)1471.

47.   Y. Shindo and R.T. Hirai, **Makromol. Chem.**, 155(1972)1.

48.   I. Harada, M. Tasumi, H. Shirakawa and S. Ikeda, **Chem.**
      Letl., 1411(1978).

49.   H. Kuzmany, **Phys. Stat. Solidi**, B97(1980)521.

50.   V. Rives-Arnau and N. Sheppard, **J. Chem. Soc.**, **Faraday**
      Trans. 1, 76(1980)394.

51.   G. Wegner, **Angew. Chem.**, 20(1981)361.

52.   R.C. Haddon and W.H. Starnes, **Polym. Preprints**, 18(1977)505.

53.   R.C. Haddon and W.H. Starnes, **Adv. Chem. Ser.**, 169(1978)333.

54.   A. Stępień and J. Wypych, **Chem. Scripta**, manuscript
      submitted.

55.   T. Yamabe, T. Matsui, K. Akagi, K. Ohzeki and H. Shirakawa,
      Mol. Cryst Liq. Cryst., 83(1982)125.

56.   P. Bernier, **Mol. Cryst. Liq. Cryst.**, 83(1982)57.

57.   F. Tüdös, T. Kelen and T.T. Nagy, in **Developments in Polymer**
      Degradation. Vol. 2, by N. Grassie, Applied Science
      Publishers, London, 1979.

58.   G. Scott, M. Tahan and J. Vyvoda, **Eur. Polym. J.**,
      14(1978)377.

114

59. V.P. Gupta and L.E.St. Pierre, **J. Polym. Sci., Polym. Chem. Ed.**, 17(1979)797.

60. D. Druesdow and C.F. Gibbs, **Nat. Bur. Stand. (US), Circ.**, 525(1953)69.

61. B. Baum and L.H. Wartman, **J. Polym. Sci.**, 45(1960)1.

62. A. Rieche, A. Grimm and H. Mucke, **Kunststoffe**, 52(1962)265.

63. M. Aldissi, M. Rolland and F. Schue, **Phys. Stat. Sol.**, A69(1982)733.

64. H.W. Gibson and J.M. Pochan, **Macromolecules**, 15(1982)242.

65. T.T. Nagy, B. Iván, B. Turcsányi, T. Kelen and F. Tüdös, **Polym. Bull.**, 3(1980)613.

66. W.H. Starnes, **Dev. Polym. Deg.**, 3(1981)135.

67. H. Kuzmany, E.A. Imhoff, D.B. Fitchen and A. Sarhangi, **Mol. Cryst. Liq. Cryst.**, 77(1981)197.

68. S.A. Liebman, D.H. Ahlstrom and C.R. Foltz, **J. Polym. Sci., Polym. Chem. Ed.**, 16(1978)3139.

69. A. Ballistreri, S. Fotti, G. Montaudo and E. Scamporrino, **J. Polym. Sci., Polym. Chem. Ed.**, 18(1980)1147.

70. W.D. Woolley, **Brit. Polym. J.**, 3(1971)186.

71. E.A. Boettner and G.L. Ball, **Pure Appl. Chem.**, 53(1981)597.

72. J.E. Brown and M.M. Birky, **J. Anal. Tox.**, 4(1980)166.

73. J.C.W. Chien, P.C. Uden and J.-L. Fan, **J. Polym. Sci., Polym. Chem. Ed.**, 20(1982)2159.

74. R.P. Lattimer and W.J. Kroenke, **J. Appl. Polym. Sci.**, 27(1982)1355.

CHAPTER 3
PHOTOLYSIS, IRRADIATION AND CHEMICAL DEGRADATION

3.1. Photolytic degradation

Comparison of the character of chemical changes in molecules excited on the electronic level with molecules whose excitement is thermally induced reveals a significant difference between the mechanism of these two reactions, and furthermore, they usually lead to entirely different products. In thermal reactions molecules in their ground state can be raised to the higher vibrational levels of the electronic ground state by collisions with other molecules or walls, and thermal energy is a factor controlling the frequency and probability of such collisions. Chemical change occurs when the particular bond accumulates sufficient energy for bond dissociation. Photochemical reactions, by contrast, involve molecules in electronically excited states. Here the excitation process is more specific for every particular chemical moiety as it depends on the quantified energy level needed to achieve molecular excitation. The energy level supplied in the photolytic process depends rather on the possibility of excitation than on its probability and frequency. The following discussion points out the differences in treatment of photolytic reactions in relation to the thermal degradation process in order to avoid unjustified comparison of changes. One is more chaotic, depending on the probability of the energy level, and the other is strictly controlled by the quantum structure of the material undergoing eventual conversion.

3.1.1. Nature of the photochemical process

Similar to the thermal degradation process, in photolytic degradation, chemical change may occur only when the molecule attains an energy level sufficient to break the weakest bond in the molecule. But it has also been known since the nineteenth century that only the light absorbed by a molecule can possibly be effective in inducing a chemical change. This last observation, known as Grotthus and Draper´s law, enforces essential limitations for the occurrence of the photolytic process. In thermal degradation, when the material temperature is low, there is still a statistical probability that some molecules will have a higher energy than others, which is sufficient to cause a chemical reaction. That is why thermally-induced reactions may proceed under a broad range of conditions, and why energy supply increases the probability of collisions and the number of molecules or bonds which qualify to undergo chemical change. Photolysis is different, since first of all energy has to be absorbed, which is not always the case. Secondly, the energy which was absorbed has to be high enough to "drive away" an electron from its former position. Let us concentrate, for the moment, on these two phenomena.

Fortunately, due to well-developed spectroscopic methods, there is a sufficient amount of information on how absorption of radiation by particular molecules occurs. The degree to which light is absorbed by matter is described by the Lambert-Beer´s law, according to which the relative decrease in the incoming beam intensity is proportional to the number of absorbing molecules encountered on the way. We can see that the intensity of the incoming beam depends, due to Planck´s equation, on the number of photons and their frequency. The number of photons removed from the beam, in turn, depends on the probability of their absorption, and this is a function of the number of absorbing atoms in unitary volume, sample thickness and, again, the probability that the photon in question will be intercepted by a given atom. It is evident that the model of absorption adapted by Lambert-Beer´s law consists of spheres of a particular cross-sectional area embedded in a transparent matrix of a known area and thickness. The probability of photon interception needs more explanation, as it determines the fate of the energy which is absorbed and, moreover, it should explain why one photon is absorbed whereas another is not. First of all, it is important to see that we have to use both of the once-conflicting theories, i.e., particulate and electromagnetic theory. The first is employed already since Planck´s equation was adopted, but if it were the only theory applied, how would we reconcile it with the fact that some photons are not absorbed? In quantum physics both theories do not exclude each other, and the dualistic character of radiation is commonly accepted. In Bohr´s model of the atom, the electron of higher energy (excited state of molecule) would be shifted to the orbit of higher energy. The electron energy difference can, therefore, be gained from radiation or released when an electron returns to its former orbit. This type of explanation is still quite common in photochemistry, regardless of the fact that Bohr´s model now has only historical value.

Taking into consideration that a light beam is composed of electric and magnetic fields, it is easy to understand why some photons can be absorbed while others cannot. Since electrons and nuclei are charged, it is not surprising that they interact with light. Each photon carries an amount of electromagnetic energy which can be calculated from its frequency. When the radiation arrives at the material, the electrons, which are otherwise in their normal state, are affected by the oscillating electric fields of electromagnetic waves. If the frequency of the light wave and molecular system do not match, the interaction is non-resonant; if they do match, the interaction is resonant. In the first case, the excess energy received by a molecule or atom is usually quite small and it is disposed with great efficiency by re-radiation. If the direction of re-radiation is different than that of the incident beam, the interaction is called light-scattering. Resonant interactions are much stronger than non-resonant ones. The characteristic frequency of absorption is actually a band of frequency of a certain width. During the act of radiation or absorption of light by matter, the electromagnetic waves constitute fields that are uniform in space but oscillating in time. These fields exert torques upon the oscillatory dipole moments of atoms and molecules. The energy transfer from electromagnetic field to matter depends on the magnitudes of the corresponding fields and dipoles, the difference in their phases and the departures of their frequences from their natural quantum frequency.

The characteristic frequency of absorption can be measured by ultraviolet spectroscopy or estimated according to Franck-Condon´s rule for molecular orbitals of the chemical

substance under consideration. Here we arrive at the
quantitative estimation of possible absorption for particular
transitions, as characterized by Table 3.1.

TABLE 3.1.

Wavelength of particular transitions.

Transition	Electron involved	Wavelength, nm
	Hydrogen	
$^2P^o-^2S$	3p-1s	102.6
$^2P^o-^2S$	2p-1s	121.6
	Carbon	
$^3P^o-^3P$	4s-2p	128.0
$^3P^o-^3P$	2p-2s	132.9
$^3P^o-^3P$	3s-2p	165.7
$^1D^o-^1D$	3d-2p	148.2
$^1P^o-^1D$	3s-2p	193.1
$^1D-^3P$	2p-2p	985.0
$^1P^o-^1S$	3d-2p	175.2
$^1P^o-^1S$	3s-2p	247.9
$^1S-^3P$	2p-2p	462.2
$^1S-^1D$	2p-2p	872.7
	Oxygen	
$^3S^o-^3P$	3s-2p	130.2
$^1D^o-^1D$	3s´-2p	115.2
$^1D-^3P$	2p-2p	630.0
$^1P^o-^1S$	3s´´-2p	121.8
$^1S-^3P$	2p-2p	297.2
$^1S-^1D$	2p-2p	557.7

Transition	Electron involved	Wavelength, nm
	Chlorine	
$^2P-^2P^o_{3/2}$	4s-3p	134.7
$^2D-P^o_{1/2}$	4s´-3p	120.1
$^2P-^2P^o_{1/2}$	4s-3p	135.1

For our purpose it is important to notice that C-Cl and C-H bonds certainly would not be affected by UV radiation because under natural conditions only light of a wavelength higher than 290nm is available, since the lower part of the light spectrum is absorbed in the atmosphere. The chromophoric groups in PVC most likely able to absorb UV radiation are the higher polyenes and the C=O group-containing compounds. Considering the absorption by particular bonds, one should remember that solvent affects absorption maxima, especially when solvent can affect the dipole moment of the molecule dissolved. Also the presence of ionized structures in the molecule affects the characteristic absorption of adjacent groups.

Photochemical reactions are closely related to changes in the oxygen ground state; therefore, it is important to discuss some details of these processes. The ground state of oxygen is a triplet state $^3\Sigma^-_g$. Molecular oxygen in the ground state has paramagnetic properties because the last two electrons, placed on antibonding orbitals $\pi^*_x 2p$ and $\pi^*_y 2p$, have parallel spin, and their total spin equals 1. The rearrangement of electron spins within these orbitals results in two singlet excited states: $^1\Delta g$ and $^1\Sigma^+_g$. The $^3\Sigma^-_g$ and $^1\Sigma^+_g$ states have the same symmetry and electron distribution, which differs from that of $^1\Delta g$ state. The lower excited state, $^1\Delta g$, cannot be produced directly from the ground state by light absorption since the transition is highly forbidden. However, it is a primary product of ozone photolysis. The energy difference between $^1\Delta g$ and $^3\Sigma^-_g$ states is low and equals 92.4 kJ/mol, compared with that of the higher excited singlet state $^1\Sigma^+_g$, which equals 159.6 kJ/mol. The energy difference for both excited states of oxygen is also considerably lower than those between the singlet ground state and the triplet excited states of most molecules. Consequently, it is highly probable that collision of $^3\Sigma^-_g$ oxygen with an excited triplet species of other molecules will produce a singlet excited state of oxygen. There is a great difference in life-time between both forms of excited oxygen, as $^1\Delta g$ has a mean life of 64.6 min, since the transition $^1\Delta g \longrightarrow {}^3\Sigma_g^-$ is strongly forbidden by the electric dipole; the mean life of $^1\Sigma_g^+$ is 12s. These values are correct in a chemically pure environment, but in the presence of many substances the mean lifetime is severely decreased due to their quenching action. The quenching rates for the $^1\Sigma_g^+$ are, in general, much faster than those for the $^1\Delta g$ state. The reactivity of singlet oxygen will be discussed below. Returning to the essence of our discussion on the nature of the photochemical process, we should still consider what happens when the molecule absorbs the energy of radiation. This is usually summarized by means of Jablonsky's diagram, which can be divided into separate stages. In the first stage, the excited singlet is formed from the molecule in the ground state:

$$S_o \xrightarrow{h\nu} S_1$$

One cannot form the excited triplet state in a similar way since direct absorption of a photon is a forbidden transition. There are also singlet states of higher energy (S_2, S_3,...), but these are very quickly converted to the lowest excited singlet state (S_1) by internal conversion; therefore, they do not usually participate in photochemical reactions. When the singlet excited state is formed, there are several possibilities for energy discharge:

$S_1 \xrightarrow{-\Delta E} S_o$	fluorescent radiation, returning to the ground state,
$S_1 \xrightarrow{-\Delta E} S_o'$	internal conversion to the ground state of higher vibrational excitation,
$S_1 \longrightarrow T_1$	intersystem crossing with formation of triplet state,
$S_1 \longrightarrow P_1 (+P_2)$	unimolecular reaction (isomerization, dissociation),
$S_1 + A \longrightarrow P_3 (+P_4)$	bimolecular reaction.

Considering a case in which the triplet state is formed, we again have several possibilities:

$$T_1 \xrightarrow{-\Delta E} S_o$$ phosphorescence with a return to ground state (possibly of higher vibrational excitation),

$$T_1 \longrightarrow S_o'$$ intersystem crossing to the ground state of higher vibrational excitation, since a triplet has lower energy than a singlet,

$$T_1 \longrightarrow P_5(+P_6)$$ unimolecular reaction (isomerization, dissociation),

$$T_1 + A \longrightarrow P_7(+P_8)$$ bimolecular reaction,

$$T_1 + B \longrightarrow S_o + B$$ excitation transfer to other molecule.

The higher triplet states (higher energy level) can be formed when the molecule in its lowest triplet state (T_1) absorbs a new photon.

From the above we can see that we have here a few groups of processes:

- radiative processes,
- radiationless processes of internal conversion and intersystem crossing,
- reactions,
- transfer of excitation.

Fluorescence and phosphorescence are the radiative processes of energy disposal for ground singlet and triplet excited states, respectively. Fluorescence is the emission of radiation in a short time (10^{-9}-10^{-5}s) as compared to phosphorescence (10^{-5}-10^{-3}s, even full seconds). The energy of radiation can still be absorbed by another molecule, but the wavelength of secondary radiation is shifted towards the red region, and is, therefore, less likely to be effective in the photolytic process. Radiationless processes are much faster than radiative ones; thus, depending on the system, they are the first to occur. Neither process contributes directly to changes in chemical structure. When the molecule absorbs the radiation energy higher than the dissociation energy of the weakest bond, there is a chance that this bond can undergo a chemical change. But whether or not this process occurs depends on the distribution of energy between vibrational, rotational and translational forms and also on the aforementioned processes and on intermolecular collisions capable of deactivating the excited state. These processes are of a strictly competitive nature, which is fortunate as otherwise only a few molecules would survive the photodegradation process.

3.1.2. Mechanism of photodegradation

Following the pattern of discussion in the previous paragraph, we would like first of all to understand if PVC can absorb the UV light available in sun-rays reaching the earth's surface, and also, which fragments of PVC structure are able to do so. Without question, if PVC contains only C-C, C-H and C-Cl bonds, it would be unable to participate in the photolytic processes, as none of these bonds is able to absorb in the range of radiation met in natural conditions, even if we estimate the lower limit at 280 nm. One serious limitation that we face in studies on the effect of UV radiation on PVC is that first of all, there are two possibilities for choosing the polymer sample. A polymer from industrial production would have one advantage - that it is similar in structure to polymers used in normal industrial practice - yet such a polymer would have numerous admixtures capable of sensitizing photodegradation by absorbing UV radiation on their own. Alternatively, polymer may be synthesized for the sake of photolytic studies, taking care that limited and controllable amounts of additives are introduced, but in this case we have a polymer whose configuration and conformation are unlike of polymers used in practical applications. More commonly, the second approach is adopted in order to deal with fewer variables. Another problem related to the specimen concerns its form, which is limited by the analytical methods used in these studies. Again, two approaches are possible: if dissolution is preferred, the polymer is analyzed in the form of solution or film-cast from such solution; alternatively, foil may be formed by a thermal process, usually on laboratory calenders. In the first case, traces of solvent and its admixtures remaining in the sample may affect the photolytic process. The second method has to introduce thermally-induced changes, which are, in fact, similar to those observed in practice, but, at the same time, are not easily quantified; therefore, they may lead to the wrong conclusions since their parameters are so numerous.

Let us now see how sample preparation method can be related to the changes observed. Recently (1), interaction between PVC and several simple ketones, such as acetone, methyl-ethyl ketone, methyl-n-propyl ketone, methyl-isopropyl ketone, diethyl ketone and diisobutyl ketone, has been studied. The C=O group IR absorption of ketones in PVC films was affected (a frequency shift was observed), which means that interaction has occurred between the ketone and PVC. The results for PVC were confirmed by observing similar frequency shifts for PVC low-molecular weight models such as 1,3-dichloropropane or n-chlorobutane. Gibb (2) and Rabek (3) observed that removing the last 2-3% of solvent in a film cast from tetrahydrofuran solution is very difficult, even when the film is dried under vacuum or treated with methanol. In other studies (4) the effect of UV radiation on the formation of tetrahydrofuran-related oxidation products was determined. Fig.3.1 shows the results. HOO-THF is not photochemically stable and photolyzes rapidly on radiation to HO-THF and butyrolactone. Fig.3.2 shows how the presence of these products affects the rate of PVC dehydrochlorination during photodegradation. The results establish that solvent effect plays an essential role in PVC photolytic studies. Also, the molecular weight of the PVC irradiated is affected in the presence of tetrahydrofuran (Fig.3.3).

When a sample is prepared instead by the thermal process, we have to include the effect of temperature on chemical changes occurring in a specimen, and furthermore, the specimen must be processed with a thermal stabilizer incorporated. We can derive conclusions in this respect from some of the data given by Bellenger (5). Fig. 3.4 shows how both temperature and stabilizer concentration (dibutyltin thioglycollate) affect absorption of UV radiation at 300 nm. The blend was prepared at 393K while extruded sample was processed in the temperature range of 453-473K. These results call for even more attention when compared with data presented in Fig. 3.5.

124

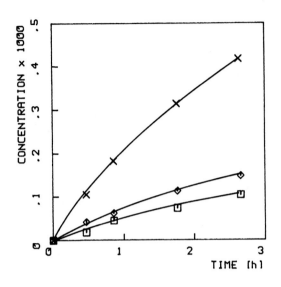

Fig.3.1. Kinetic curves of tetrahydrofuran oxidation products formation ✕ - HOO-THF, ◇ - HO-THF, ☐ - butyrolactone during irradiation at 254 nm. (Modified from Ref. 4.)

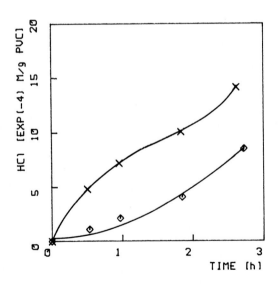

Fig.3.2. PVC dehydrochlorination during UV irradiation at 254 nm. ◇ - PVC, ✕ - PVC + HOO-THF (1%). (Modified from Ref. 4.)

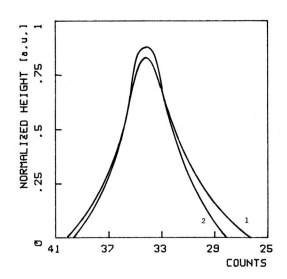

Fig.3.3. Molecular weight distribution of PVC after UV irradiation for 3 hrs at 254 nm. 1 - PVC, 2 - PVC + THF (2%). (Modified from Ref. 4.)

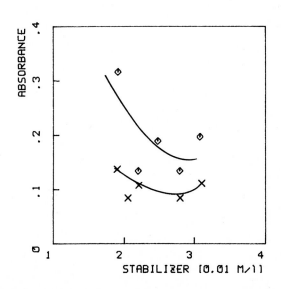

Fig.3.4. UV absorbance at 300 nm. x - blends, ◇ - extrudates. (Modified from Ref. 5.)

126

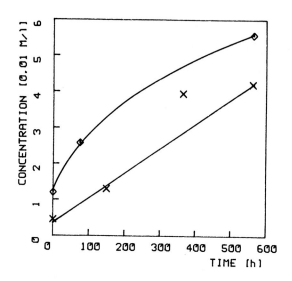

Fig.3.5. Concentration of total chlorides against irradiation time. × - blends, ◇ - extrudates. (Modified from Ref. 5.)

Similar conclusions can be reached from reviewing other related studies (6-9).

Torikai (10) gives the UV absorption of PVC films prepared from tetrahydrofuran solution in initial form and after irradiation (Figs. 3.6 and 3.7). Evidently, the film is able to absorb UV in the range of 250-290nm even before irradiation. Salovey (11) attributes this absorption band to the presence of two radicals:

$$-CH_2-CH=CH-\overset{\bullet}{C}H-CH_2-CHCl-$$

$$-CH_2-(CH=CH)_n-\overset{\bullet}{C}H- \quad (n = 2-4)$$

Absorption at 410 and 620 nm is due to the dehydrochlorination reaction in which polyenyl radicals are formed:

$$-CH_2-(CH=CH)_n-\overset{\bullet}{C}H-$$

The concentration of radicals depends on temperature, as seen in Fig.3.8.

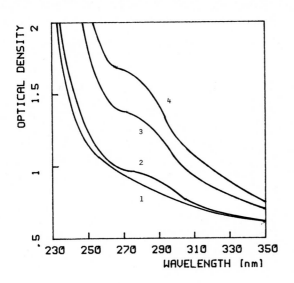

Fig.3.6. Optical absorption spectra of PVC photoirradiated film
at 313K. 1 - control, 2 - 10 min (irradiation), 3 - 30 min, 4 -
50 min. (Modified from Ref. 10.)

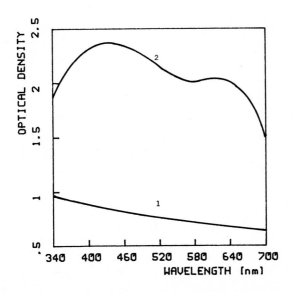

Fig.3.7. Optical absorption spectra of PVC photoirradiated film
at 373K. 1 - control, 2 - 30 min of UV irradiation. (Modified
from Ref. 10.)

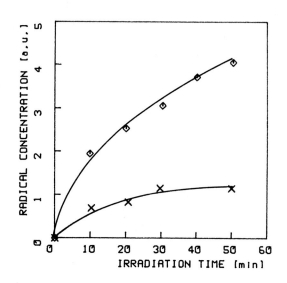

Fig.3.8. Radical concentration versus irradiation time at different temperatures. × - 326K, ◇ - 373K. (Modified from Ref. 10.)

Apparently, the radical decay rate depends on the temperature, and it is increased above T_g. According to Rånby (12), the concentration of alkyl radical above 323K and polyenyl radical above 413K is around the detection limit. Also, the atmosphere in which irradiation is performed affects radical concentration (Fig. 3.9).

The presence of air on irradiation causes oxidation of the polyene radical and formation of a peroxy radical:

$$-\overset{|}{\underset{|}{C}}OO^{\,\cdot}$$

a finding confirmed by ESR spectroscopy. Formation of these radicals and carbonyl groups is also known from other studies (13, 14) by IR spectrophotometry. The concentration of CO groups depends on film thickness (Figs. 3.10 and 3.11.)

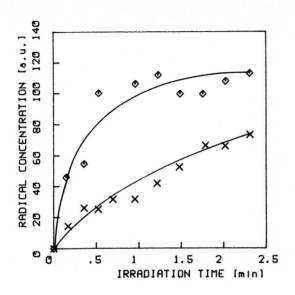

Fig.3.9. Radical concentration versus irradiation time in vacuum
(◇) and air (×). (Modified from Ref. 10.)

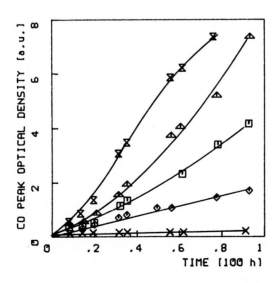

Fig.3.10. Optical density of CO peak versus irradiation time for
films of varying thickness: × - 10 μm, ◇ - 38 μm, □ - 73 μm, Δ -
117 μm, ⊠ - 175 μm. (Modified from Ref. 14.)

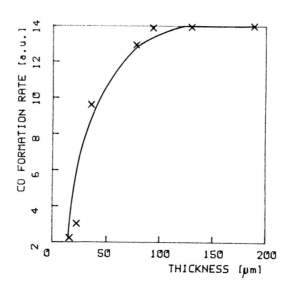

Fig.3.11. Maximal rate of C=O group formation versus thickness. (Modified from Ref. 14.)

The results, as commented on by one author (14), show that the presence of HCl is a predominant factor in CO group formation, since as temperature increases, the number of CO groups decreases. The HCl diffusion increases as temperature rises (Fig.3.12). Verdu (14) believes that there is a critical temperature around 318K above which the accumulation of HCl in a sample is no longer possible, and thus the autocatalysis effect decreases. Accumulation of the CO group in the polymer increases absorption of UV radiation, as depicted below (Fig.3.13). This explains why, along with the photodegradation process, its effectiveness increases, as further absorption is possible; therefore, more energy can be utilized for the photolytic process (Fig.3.14).

Thermal processing may also introduce peroxy-groups and therefore enhance the photooxidation rate, as is evident from Figs. 3.15 and 3.16 (16).

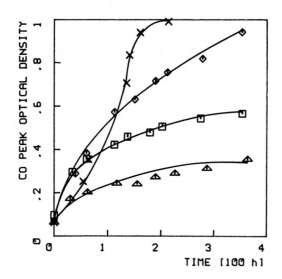

Fig.3.12. Optical density of CO peak versus irradiation time for varying temperature. × - 308K, ◇ - 318K, □ - 334K, △ - 344K. (Modified from Ref. 14.)

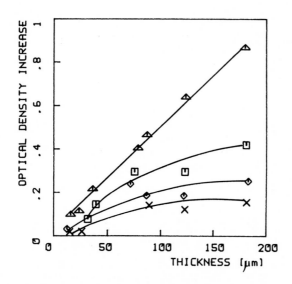

Fig.3.13. Optical density increase versus sample thickness after 765 hrs of UV irradiation at varying wavelength. △ - 275 nm, □ - 300 nm, ◇ - 320 nm, × - 350 nm. (Modified from Ref. 14.)

132

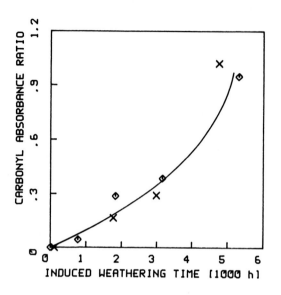

Fig.3.14. Change in carbonyl group concentration during natural (×) and induced (◇) weathering. Six months of natural weathering equals approximately 1,500 hrs of induced weathering on the time scale. (Modified from Ref. 15.)

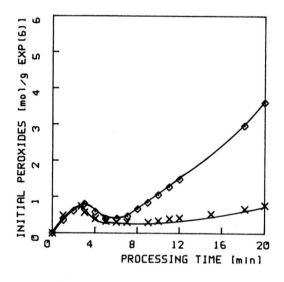

Fig.3.15. The effect of processing time on initial peroxide concentration. × - 443K, ◇ - 483K. (Modified from Ref. 16.)

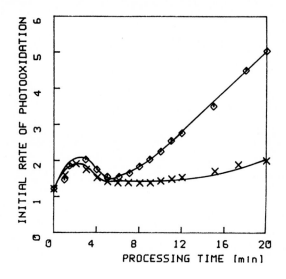

Fig.3.16. The effect of processing time on the initial rate of photooxidation. × - 443K, ◇ - 483K. (Modified from Ref. 16.)

Taken together, Figs. 3.13 , 3.15 and 3.16 reveal how photodegradation is initiated: Thermally induced chemical changes in the material allow for a means of UV absorption; hence, photochemical mechanisms are able to perform. Nonetheless, many investigators are not satisfied with such an explanation alone. They argue that some small amount of admixtures in PVC must be present to absorb the first quants of ultraviolet energy and thus sensitize photochemical reactions. There is no direct evidence due to the isolation of such species or to comparison of material photolysis rate with admixtures after they have been extracted from the material. The assumption is based mainly on knowledge of the technological process of PVC manufacturing and the presence of possible additives. Some investigators attempt to show photodegradation from this point of view. Metals such as Co, Cu and Fe are singled out as possible offenders (17) due to their own absorption in the UV range and their catalytic effect on hydroperoxide decomposition. P-cresol and benzophenone are the most frequently mentioned sensitizers (18, 19). Heller (18) and Hirayama (19) show that both chemical substances are effective photosensitizers of PVC photodegradation, but there is no direct answer as to how they can be incorporated into the PVC polymer. One photosensitizer, a phenolic type antioxidant included in PVC thermal stabilizer, was identified by Foster (20). It seems that either inclusion or exclusion of sensitizers from the process of PVC photolysis is not supported by available data, because on one hand it is well proved that absorption of UV light by thermally degraded PVC is possible, and in fact, PVC in practical use has always been degraded thermally before being exposed to weathering. The fact that some antioxidants contained in thermal stabilizers may play a role in sensitizing is largely irrelevant because antioxidants were deleted from PVC stabilizing mixtures long ago, well before Foster (20) published his paper. On the other hand, even traces of substances formed by side-reactions can be efficient in photosensitizing because they only start a photolytic reaction that can later develop according to the above mechanisms.

We have thus far discussed conditions sine qua non for photochemical reaction, but UV absorption alone does not ensure that a photochemical reaction takes place. We should be able to demonstrate that certain thermodynamic conditions must exist for such absorption to be effective, and in order to prove the mechanism of photodegradation, we also should be able to identify the products of particular stages of the photochemical process. In doing so we can possibly predict the direction of changes occurring in the polymer.

Based on the energy of monochromatic light corresponding to bond dissociation energy, we have some data for low molecular models as given in Table 3.2.

The trend for this data is quite obvious:

1. practically every bond from irregular segments of the PVC chain can be affected by the energy level available in sun-rays,

2. the more electron-withdrawing group adjacent to the bond, the higher probability of bond dissociation.

TABLE 3.2.

The wavelength of light of energy corresponding to the bond
dissociation energy for various compounds.

(Modified from Ref. 21.)

Structure	Wavelength, nm
$CH_3-\overset{\cdot}{\underset{H}{C}}H-CH_3$	302
$CH_3-\overset{\cdot}{\underset{H}{C}}Cl-CH_3$	308
$CH_3-\overset{CH_3}{\underset{H}{\overset{\mid}{C}}}-\overset{\cdot}{C}H_3$	315
$CH_2Cl-\overset{CH_3}{\underset{H}{\overset{\mid}{C}}}-\overset{\cdot}{C}H_2Cl$	319
$CH_2=CH-CH_2 \dagger H$	369
$CH_2=CH-CHCl \dagger H$	379
$CH_2=CH \dagger H$	232
$CH_3-\overset{\cdot}{\underset{Cl}{C}}H-CH_3$	389
$CH_3-\overset{CH_3}{\underset{Cl}{\overset{\mid}{C}}}-\overset{\cdot}{C}H_3$	412
$CH_2=CH-CH_2 \dagger Cl$	423
$CH_2=CH \dagger Cl$	232
$CH_3 \dagger CH_3$	337
$CH_2Cl \dagger CH_3$	346
$CH_2=CH-CH_2 \dagger CH_3$	460
$CH_2=CH \dagger CH_3$	260

Fortunately for polymer stability, the real effect does not
strictly obey the energy rule, since a substantial part of the
energy supplied is always wasted by ineffective rotational and
vibrational motion; therefore, the energy supply level should be
well above the lowest limit given in Table 3.2, if the bond is

meant to dissociate. One important feature which used to be observed in natural and artificial weathering studies (15, 22) is that the sample temperature during irradiation also plays an essential role. The sample temperature explains why Weather-O-Meter is more effective than Xenotest in photolytical changes. It does not seem possible that such low temperatures can affect changes on much higher energy levels, but the temperatures are not actually so low. In fact, it is not unusual that during natural weathering a sample has a temperature around 373K, and, in artificial weathering tests, a temperature above 333K is common. This energy level, then, can already be utilized for rotational-vibrational motion, with the result that on photon absorption, a higher portion of energy can be used for translational motion, thus causing more efficient photolysis.

Mori (13) compared photolysis of 2,3-dichlorobutane and 2,4-dichloropentane, i.e., low molecular models of head-to-head and head-to-tail structures, respectively. His results showed that head-to-head structures are not likely to be initiation centers in photolytic degradation. In other studies on model compound (18), sec-butylchloride and t-butylchloride were used as a model for unbranched and branched chains, respectively. The photolytic reaction sensitized by benzophenone yielded the following products:

$$CH_3-\underset{\underset{CH_3}{|}}{\overset{\overset{CH_3}{|}}{C}}-Cl \xrightarrow{h\nu} \underset{CH_3}{\overset{CH_3}{\diagdown}}C=CH_2$$

$$CH_3-CH_2-\underset{\overset{|}{CH_3}}{\overset{CH_3}{\overset{|}{C}}}HCl \xrightarrow{h\nu} CH_3-CH=CH-CH_3 + CH_3-CH_2-CH=CH_2$$

Hiroyama (18) and Harriman (23) proved that a benzophenone sensitization reaction is a triplet state reaction similar to that of p-cresol.

ESR spectroscopic studies on both model compounds and polymer have yielded invaluable information on the radicals obtained directly after photons are absorbed. According to earlier predictions (24), the following radical was expected to be formed:

$$-CH-CH_2-\overset{\cdot}{C}H-CH_2-CH-$$
$$\phantom{-CH-CH_2-\overset{\cdot}{C}H-}||$$
$$\phantom{-CH-CH_2-\overset{\cdot}{C}H-}ClCl$$

This radical, if present, should have a six-line spectrum, but such anticipated ESR absorption was not experimentally established. Recently, Yang (25) understood that the measuring problem prevented this prediction from being confirmed. When measurement was done at liquid nitrogen temperature, a six-line spectrum for both polymer and 3-chloropentane was obtained without difficulty. If the temperature is raised to 163K, the radical will produce a five-line spectrum (the most frequently measured before), which is typical for a radical of the following chemical structure:

$$-CH-CH_2-\overset{\cdot}{C}-CH_2-CH-$$
$$|||$$
$$ClClCl$$

Further heating to 298K produces a polyenyl radical:

$$-(CH=CH)_n-\overset{\cdot}{C}H-$$

From this we can see how changes gradually occur until we reach the final stage of a polyenyl radical observed in conditions of photodegradation.

Let us now summarize the above in order to predict the sequence of chemical reactions occurring during the photolytic process of PVC degradation. As some of the observations are not fully confirmed, we should show the process alternatively to allow for further clarification still to come. The first stage includes initiation of the process, and can be written in the following manner:

(a) Radical formation by impurity (I) sensitization (10):

$$PVC(I) \xrightarrow{h\nu} PVC(I)^*$$
$$PVC(I)^* \longrightarrow -CH_2-\overset{\cdot}{C}H-CH_2- + Cl^{\cdot}$$

(b) Benzophenone as impurity (19):

$$(C_6H_5)_2C=O \xrightarrow{h\nu} {}^1(C_6H_5)_2C=O \longrightarrow {}^3(C_6H_5)_2C=O$$

(c) Tetrahydrofuran-related products as an impurity (4):

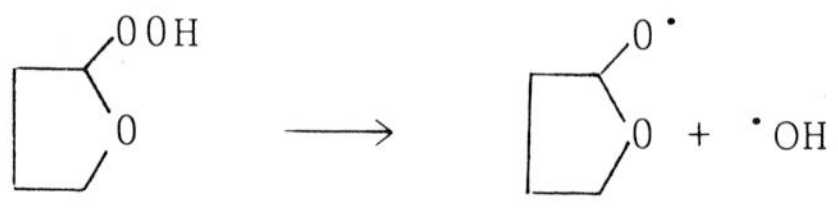

(d) Molecular-singlet oxygen initiation (26):

$$-(CH=CH)-_n \xrightarrow{h\nu} -(CH=CH)^*-$$

$$-(CH=CH)^*- + {}^3O_2 \longrightarrow -(CH=CH)- + {}^1O_2({}^1\Delta g)$$

$$-(CH=CH)- + {}^1O_2 \longrightarrow \underset{\underset{OOH}{|}}{-CH-CH=CH-}$$

(e) Carbonyl-group-in-chain initiation (12):

Further steps lead to the double-bond formation, and we can write them according to the scheme proposed by Yang (25) quoted above or as follows:

$$-\underset{\underset{Cl}{|}}{CH_2-CH-} + R^\bullet \xrightarrow[-RH]{} -\underset{\underset{Cl}{|}}{\overset{\bullet}{C}H-CH-} \longrightarrow -CH=CH- + Cl^\bullet$$

Cl$^\bullet$ radical can still participate in further reaction:

$$-\underset{\underset{Cl}{|}}{CH_2-CH-} + Cl^\bullet \longrightarrow -\underset{\underset{Cl}{|}}{\overset{\bullet}{C}H-CH-} + HCl$$

The participation of HCl in UV photolysis can be explained by **migration** of short polyene sequences along the polyene chains and formation of longer ones (27). Verdu (14) proposed a mechanism whereby HCl can be substituted into the double bond on UV irradiation, but this seems a less realistic explanation of the photobleaching phenomenon. The crosslinking reaction is still not fully understood. Ranby (12) explains it by intermolecular electron delocalization between two polyene structures. This explanation, however, is a matter of speculation rather than experiment, because there is no way to measure crosslinks, and actually no method is available so far that can give reliable results in this respect. Further dehydrochlorination leading to the propagation of polyene chains can be described in a way similar to the initiation process shown above in chemical equations. We can see recent interest and better understanding of the photolytic process, but many questions still remain unanswered. The limitations of our knowledge are evident from the protection methods used in practice, since the only method of PVC protection at the moment is based on the so-called screening effect of some organic substances called UV absorbers or inorganic fillers, which play a similar role. The problem is that these UV absorbers are themselves not sufficiently resistant and are also degraded in the process. So far it has not been possible to find for PVC, unlike for some other polymers, substances able to quench and deactivate excited states which are the most efficient in coping with photodegradation process.

Let us now review the effect of UV irradiation on changes characteristic for PVC. The molecular weight distribution before and after irradiation is given by Fig.3.17. This data can also be supported by Rabek´s results (26), which indicate that chain-scission depends on the atmosphere in which irradiation is performed.

Matsumoto (15) has compared the molecular weight and molecular weight distribution change during natural and induced weathering. His data are given in Figs. 3.18 and 3.19.

Figs.3.8-3.16 reveal how radicals formed during photolysis are converted to the C=O group and how this process affects polymer stability in photolytic reactions. Fig.3.20 shows the extent of the same effect on the thermal degradation of polymer, which is an essential factor if one were to apply partially degraded product for re-cycling (28).

140

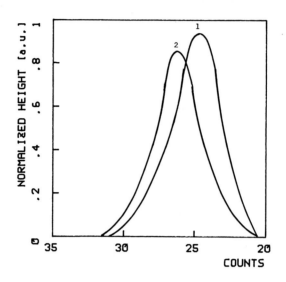

Fig.3.17. Molecular weight distribution of PVC. 1 - control, 2 - irradiated for 100 hrs at 273K. (Modified from Ref. 13.)

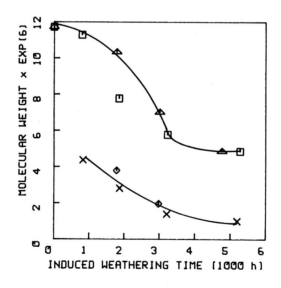

Fig.3.18. Change in the molecular weight during natural ($\triangle,\diamond$) and induced ($\times$,$\square$) weathering. Six months of natural weathering equals approximately 1,500 hrs of induced weathering on the time scale. (Modified from Ref. 15.)

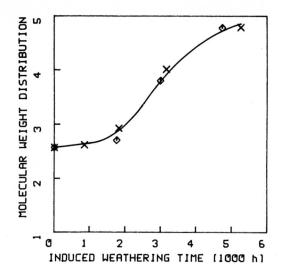

Fig.3.19. Change in molecular weight distribution during natural (◇) and induced (×) weathering. Six months of natural weathering equals approximately 1,500 hrs of induced weathering on the time scale. (Modified from Ref. 15.)

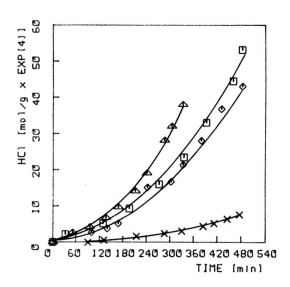

Fig.3.20. The effect of UV irradiation on thermal dehydrochlorination of PVC. Degradation in nitrogen at 473K. × - control, ◇ - 1 hr (UV irradiation), □ - 2.25 hrs, △ - 3 hrs. (Modified from Ref. 28.)

3.1.3. Double bond photodegradation in thermally degraded PVC (photobleaching)

When β-carotene-doped PVC films are irradiated with a UV-lamp having maximum emission at 365nm, the color of the sample changes as seen in Fig. 3.21 (29).

Similar changes account for PVC films formerly being thermo-degraded (Fig. 3.22). This phenomenon, called photobleaching, was studied by Owen (27, 30, 31).

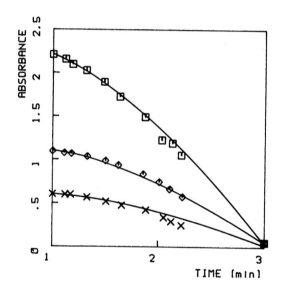

Fig.3.21. Absorbance at 450nm of PVC film containing varying amount of β-carotene versus irradiation time in presence of oxygen. × - 0.5%, ◇ - 1%, □ - 2%. (Modified from Ref. 29.)

Why does polymer color increase during the normal photolytic process while under certain circumstances it disappears almost completely? Kohn (29) suggested that the HCl elimination reaction occurs only on radiation below 340 nm while in photobleaching studies care was taken to limit the radiation range as much as possible. These studies could have given important answers, but they faced problems encountered in other photolytic studies, e.g., the poorly established effect of solvent presence since measurements were carried out in solutions or films which were obtained from these solutions. At the moment there is confusion about the effect alone without including the additional influence of changes attributable to solvent oxidation products.

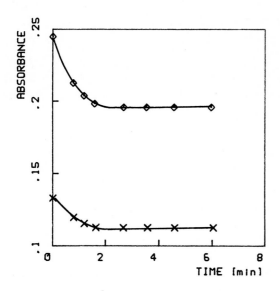

Fig.3.22. Evolution of the 312nm absorbance during photooxidation of two suspension PVC types in form of films. × - PVC polymerized in absence of oxygen, ◇ - PVC polymerized in presence of oxygen. (Modified from Ref. 30.)

In addition to the formation of carbonyl groups, formation of cyclohexadiene structures is theoretically possible (31):

If this were the case, polyene sequences would be shortened as double bonds in the cyclic moiety are not a part of the conjugated system. Actually, the conjugated system above has lost two double bonds. The above assumption is not fully supported by experiment, and therefore is best regarded as speculation. Furthermore, HCl may participate in photobleaching reactions and may eventually be added to double bonds (30, 31). If this is the case, how it is possible that we are reaching horizontal asymptote, as on Fig.3.22, and photobleaching, on one hand, and dehydrochlorination during UV irradiation on the other? If these changes are due to the reversible nature of HCl formation-addition reactions, we should not be able to obtain photobleached polymer when HCl is removed from the system. We must admit that current data for PVC cannot explain what really happens to polyenes when they contact oxygen with and without UV irradiation.

One can try to find an analogy from the studies on polyacetylene (32). In the absence of UV light, the oxidation of polyacetylene involves triplet oxygen, which can interact only with existing free radicals. In the presence of UV light, oxidation is most likely complicated by the presence of excited species. Three reactions are possible:

$$\text{excited polyene-}^{3}O_{2}$$
$$\text{excited polyene-}^{1}O_{2}$$
$$\text{ground state polyene-}^{1}O_{2}$$

It is known that electric conductivity is increased at the initial stage of oxidation and then irreversibly decreased, probably because of the formation of carbonyl groups and the decomposition of the polymer chain. Also, mechanical properties change on oxidation, and the result is products of lower tensile strength and elasticity. The results obtained for polyacetylenes and those presented above for PVC differ in one important respect. From Figs. 3.10 and 3.11 we can see that as PVC film thickness increases, the CO group concentration also increases (14). The observations for polyacetylene are just reversed as thick films oxidize more slowly on irradiation since UV light penetrates a smaller fraction of thicker films. These differences are attributable either to Verdu´s (14) inability to include the effect of sample thickness on optical density of the CO group, which otherwise would show a trend similar to that of polyacetylene, or more likely, these differences support Verdu´s explanation that the HCl diffusion rate is a factor controlling carbonyl group formation.

There is still a need to find data that can either support or reject some of the above suggestions. Especially we should expect a clear answer concerning the composition of products of the photobleaching process and possibly be able to relate this composition to the wavelength of light used for sample irradiation. Such information would improve our understanding of material changes during more advanced stages of natural weathering.

3.2. Irradiation of poly(vinyl chloride)
3.2.1. Energy of irradiation in comparison to thermal degradation and photolysis

From comparing the effect of thermal degradation and UV irradiation, we can say that photolysis can be placed between thermal degradation and pyrolysis. The energy level supplied in both thermal and UV degradation is on a comparable level, which is not the case in high energy radiation, e.g., γ-rays, in which the energy level is far higher. UV radiation in the range of 290 to 380 nm has quant energy in the range of 4.3 to 3.3 eV, whereas the average energy of quants of γ-rays emitted by ^{60}Co equals 1.25 MeV. Comparison of these values shows how great an energy disparity exists between UV and high energy radiation. The energy of γ-rays is higher than the energy necessary for any bond to be dissociated; they can directly cause formation of double ions, which is not possible for UV rays. The mechanism of quant absorption is different here than in UV radiation. In the latter only specific radiation can be absorbed; therefore, absorption is limited by polymer structure. In case of γ-radiation the situation is different, since radiation energy exceeds the order of magnitude of bond energy and ionization energy. Thus, absorption depends only on material density, i.e., the number of atoms in material volume. Also, in case of high-energy radiation, ability to penetrate the material is its distinctive feature. UV photolysis is concentrated mainly on the material surface, while high-energy radiation acts on the full material volume. From this brief comparison, we can see that the changes expected may vary from those caused by thermal degradation and UV photolysis.

3.2.2. Mechanism of the process

Similar to other decomposition processes, the dehydrochlorination reaction is also typical for the high energy radiation degradation process. ESR spectrum of γ-irradiated PVC sample contains a septet line-spectrum attributed to the following radical (33):

$$-CH-CH_2-\overset{\bullet}{C}H-CH=CH-CH_2-CH-$$
$$\quad\ \ |\qquad\qquad\qquad\qquad\qquad |$$
$$\quad\ \ Cl\qquad\qquad\qquad\qquad\quad Cl$$

which at higher temperatures is converted to a polyene radical (34). Irradiated PVC generates HCl continuously during storage at 353K (35). Samples of PVC reduced with tri-n-butyltin hydride, according to the procedure of Starnes (36), when irradiated with ^{60}Co γ-rays, produced low molecular weight alkanes and alkenes (37). This may indicate that branches react first under γ-irradiation. Under inert-gas atmosphere, alkyl radicals are formed (12), which obviously must lead to a decrease in the molecular weight of the polymer. The radicals are transformed to peroxyradicals (38, 39) when a sample, on irradiation, is exposed to air. It is interesting that elongated (drawn) films exhibited an increase in carbonyl index after γ-irradiation, which was explained by better oxygen solubility in drawn PVC films.

Irradiation in the presence of air causes more chain scissions (40) than that in inert atmosphere, which facilitates crosslinking (41). Crystalline order in PVC is destroyed irradiation (38). Koch (42) applied γ-radiolysis to monitor the lauric acid from thermal stabilizer substitution into the PVC chain by measuring undecane evolvement. Recent papers (43-45), have reported that epoxy stabilizers act on γ-irradiation based on a similar mechanism to that in thermal degradation. The main difference is in the number of substitutions into the chain.

From this short review of high-energy irradiation of PVC, it is evident that the processes are not of great interest, since there is little practical application of this irradiation process. At the moment, there are two reasons to study these properties. One is that some materials wrapped in PVC foil can be radiation sterilized, in which case we should expect changes in the mechanical properties of the material and we might be interested in the composition of radiolysis products, especially as some radicals are a very long-lived species (12). The other possibility is to use irradiation to destroy waste polymeric products, which was intensively investigated in a paper in Guillet's book (38). His conclusion is that, due to initial irradiation, one can decrease the temperature of further decomposition to 423K. The maximal effect of radiolysis is achieved at lower doses so far as degradation of waste products is concerned. Generally, this field does not have much relevance at the moment, especially as changes recorded on high energy irradiation are not able to explain other decomposition processes, since the level of energy applied is not comparable with other processes used in practice.

3.3. Chemical dehydrochlorination of polymer

After the first studies on chemical dehydrochlorination had been published (46), no further interest was shown in that field for about 20 years. At the beginning of the 1960's some work was initiated on chemical dehydrochlorination in order to obtain semiconductive polymers that could eventually be used in modern electronics (47, 48). Studies in that field have been partially given up as new possibilities were shown to be more attractive - e.g., polyacetylene synthesis and especially compounds containing metal ligands that can form one-dimensional polymers with a metal core. But some papers are still published on analysis of thermally and chemically degraded PVC in respect of electrical conductivity.

At present there are several areas in which studies on the chemical dehydrochlorination of PVC have continued. Recent research has studied the kinetics of reaction (49-56); compared thermal and chemical dehydrochlorination (31, 57); compared the changes in PVC physical properties after chemical, thermal and photochemical degradation; and correlated alkaline dehydrochlorination kinetics with the length of polyenes formed

(31, 57).

Some theoretical issues are also connected with these studies, such as preparation of material for analysis of the physical behavior of semiflexible polymer in solution in order to determine the structure of a polymer chain (52). Other work done on the interaction of chemically dehydrochlorinated polymer and trifluoroacetic acid was aimed at studying the possibility of the migration of double bonds along the polymer chain (58, 59). Additional work has been done on the preparation of raw material for further chemical modification (60) and chlorination (61) or block copolymer synthesis (62). Recently, a good deal of interest has been devoted to the so-called photobleaching reaction (4, 19, 26, 29, 31, 57) in order to understand the mechanism of that reaction, the contribution of polyene sequences on absorption in UV and visible regions, factors limiting color formation in degraded PVC, and elaboration of the methods of measuring the concentration of short polyenes present in initial polymer.

3.3.1. The methods of chemical dehydrochlorination of PVC

The procedure of chemical elimination of HCl from the PVC chain developed by Benough (50) and Shindo (49), after a few modifications, is used at present. Poly(vinyl chloride) is dissolved or swollen under an oxygen-free atmosphere. Care is always taken that solvents are peroxide-free (purified prior to use) in order to avoid oxydation of the double bond formed in the elimination reaction. After the polymer has been dissolved, the ethanol solution is added to initiate the reaction. Typical experimental conditions are shown in Table 3.3 (56). The reaction is followed by neutralization with diluted HCl; then the product is left in distilled water for 24 hrs, dried in a vacuum over P_2O_5 and stored under nitrogen in darkness. Taking into consideration the reaction conditions, one should expect the possibility of interaction of solvent traces left in the sample with PVC molecules (29) and formation of charge transfer complexes with THF (63), which might influence the results of some studies performed on these materials.

TABLE 3.3.

Reaction conditions for alkaline dehydrochlorination of PVC.

(Data from Ref. 56.)

Solvent type	Solvent ml.	PVC g	Ethanol ml.	KOH g	Temp. K	Reaction time, min
THF	550	20	50	5	278	300
THF	1200	150	250	26.5	280	-
Dioxan	360	127	75	8	288	-
Dioxan	900	310	150	16	293	480

3.3.2. Kinetics and mechanisms of reaction

Reaction kinetics can be studied by titration of the unreacted KOH or KCl produced from reaction or by UV-visible spectroscopy. Flodin et al. (55, 56) used automatic base-acid titration and mercuric nitrate titration to establish the HCl elimination rate (Fig. 3.23).

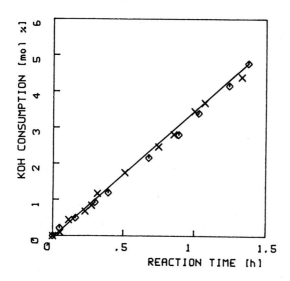

Fig.3.23. The degree of dehydrochlorination versus reaction time: (×) from KOH consumption; (◇) from Cl evolution. (Modified from Ref. 55.)

The reaction rate remains constant, at least if not more than 30% of the KOH introduced was reacted, as was the case here. The reaction followed a pattern of pure elimination, as none of the hydroxyl groups was detected by IR spectroscopy.

The concentration of reagents determines the rate of dehydrochlorination, as illustrated in Fig. 3.24.

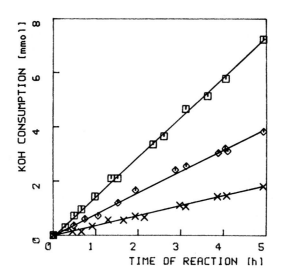

Fig.3.24. KOH conversion at 282.5K versus reaction time for various compositions. $\times$ - 20.2g PVC + 9.62 mmole KOH, $\diamondsuit$ - 20g PVC + 29.87 mmole KOH, $\square$ - 40.7g PVC + 31.14 mmole KOH. (Modified from Ref. 55.)

As an increase in concentration of any reagent causes a more or less proportional increase in reaction rate, the reaction looks like a second-order reaction. The same reaction order was proposed by Shindo, who has shown the process of HCl alkaline elimination as the chain of consecutive reactions which follows:

$$KOH + PVC \xrightarrow{k_1} P_1 + KCl + H_2O$$
$$KOH + P_1 \xrightarrow{k_2} P_2 + KCl + H_2O$$
.
.
.
$$KOH + P_{n-1} \xrightarrow{k_n} P_n + KCl + H_2O$$

which rates form the set of differential equations:

$$d[P_1]/dt = k_1[KOH][PVC] - k_2[KOH][P_1]$$
$$d[P_2]/dt = k_2[KOH][P_1] - k_3[KOH][P_2]$$
.
.
.
$$d[P_n]/dt = k_n[KOH][P_n]$$

Flodin assumed that concentrations of partially reacted products are small when compared with PVC concentration and arrived at a general equation, as follows:

$$d[E]/dt = k_1[KOH][PVC]$$

where [E] denotes the total concentration of double bonds in all polyene sequences. Having this equation, we can see that no comparison can be made between the dehydrochlorination reaction under elevated temperature and the reaction caused by alkaline elimination. From the point of view of molecular collision dynamics, the reaction rate depends, among other parameters, on the size of the molecules, the collision energy (both values are included in a so-called collision cross-section) and the energy of chemical change. All these values are entirely different for both cases, especially because in alkaline elimination we are dealing with a small molecule of KOH, which is the component of bimolecular reaction.

Additional data of interest on alkaline dehydrochlorination are found in the work of Shindo (52), who monitored the unsaturation formation rate in the reaction course by UV spectroscopy. Fig. 3.25 shows how the molecular weight of polymer determines the rate of double bond formation, depending on reaction time.

The elimination rate increases with molecular weight, which differs substantially from the results obtained for thermal degradation of PVC. For the discussion of results, Shindo derived the following equation:

$$\frac{[\eta]}{[\eta_0]} = \left[\frac{<R^2>}{<R_0^2>}\right]^{3/2}\left[1-0.584\sum_i^{\rho} n_{2i}/D\right]^{-1}$$

in which:

$$\frac{<R^2>}{<R_0^2>} = \frac{\sigma'^2}{\sigma^2} - \frac{\sigma'^2}{\sigma^2}\cdot\frac{\sum_i^{\rho}n_{2i}}{D} + \frac{a^2}{l^2}\frac{1-\cos\theta}{1+\cos\theta}\frac{\sum_i^{\rho}n_{2i}}{\sigma^2 2D}$$

where:

$<R^2>$ - the mean-square end-to-end distance of the dehydrochlorinated chain,

$<R_0^2>$ - the mean-square end-to-end distance of the initial chain,

σ - factor due to the hindrance of internal rotations (for PVC=1.83),

σ' - factor due to the hindrance of internal rotations in dehydrochlorinated chain,

n_{2i} - number of conjugated double bonds in sequence,

D - degree of polymerization,

a - length per conjugated double bond unit,

l - bond length,

θ - supplement of the bond angles,

$[\eta]$ - intrinsic viscosity of initial polymer,

$[\eta_0]$ - intrinsic viscosity of dehydrochlorinated polymer. Based on this relationship, the ratio $[\eta]/[\eta_0]$ has been calculated and compared with values of that ratio obtained from experiment, then plotted versus the percentage of conjugated double bonds in the dehydrochlorinated chain as shown in Fig. 3.26.

The data sets from both methods (titration and UV spectroscopy) cannot be correlated successfully, especially for early stages. This suggests the need for further work, especially that UV spectroscopy is used for measuring the distribution of polyene sequences in PVC thermal degradation studies.

Recently (29) the following polyene length distribution and physical properties have been proposed for PVC extrudate as shown in Table 3.4.

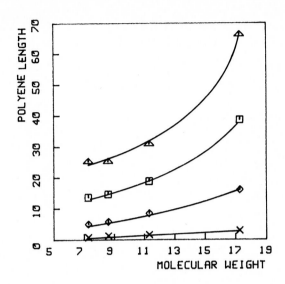

Fig.3.25. Molecular weight versus the total number of conjugated polyene sequences in a dehydrochlorinated PVC chain at various reaction times. x - 2 hrs, ◇ - 4, □ - 6, △ - 8. (Modified from Ref. 52.)

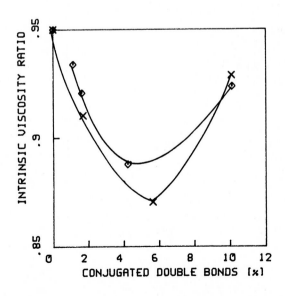

Fig.3.26. Intrinsic viscosity ratio versus percentage of conjugated double bonds in dehydrochlorinated PVC. x - experimental, ◇ - theoretical. (Modified from Ref. 52.)

TABLE 3.4.

Polyene length distribution in PVC extrudate.

(Data from Ref. 29.)

n	λ_{max}, nm	ε_{max}, 1000/mol cm
4	312	73
5	329	121
6	345	138
7	371	174
8	395	204
9	421	233
10	445	261
11	467	292
12	485	320

The values of λ_{max} vary in different papers (29,31, 56). The typical UV spectrum of a sample degraded by alkaline elimination is shown in Fig. 3.27.

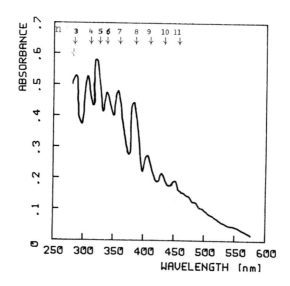

Fig.3.27. UV-visible spectra of PVC sample after KOH elimination reaction in THF. (Modified from Ref. 56.)

Evidently, isolated double bonds and diene sequences cannot be seen from the spectra, and trienes can be detected only with difficulty.

Wiersen and Flodin (56) did not observe the crosslinking reaction in their experiment on alkaline elimination of HCl. The only reason for molecular weight change was chain scission, the rate of which could be restricted to one or below per polymer chain with samples of low conversion.

Ozonolysis was performed to determine the length of polyene sequences, followed by determination of molecular weight distribution. Comparing the molecular weight distribution of the initial polymer to that being dehydrochlorinated and subjected to ozonolysis, Wiersen and Flodin (56) concluded that active centers in which the elimination reaction was initiated were randomly distributed. Hence, the reaction started in some weak points of the polymer chain and continued to produce longer sequences in the further course of the process. The places of initiation were distributed at an average distance of 150 carbon atoms for suspension polymer, and the product of the elimination reaction could be described by the general formula:

$$-\{(CH_2-CHCl)_{70}-(CH=CH)-_{1-14}\}_m^-$$

The investigators believed that only a few sequences had more than 15 double bonds, but the data presented and the methods used cannot guarantee that such a belief is valid, as practically speaking, more than 11 bonds in conjugation cannot be detected on the UV spectrum.

The mechanism of the elimination reaction is still unresolved. Wiersen and Flodin (56) concluded that dehydrohalogenation of alkyl halides resembles the E2 mechanism, but their explanation seems to be insufficient for two main reasons. First of all, a reaction performed according to the E2 mechanism could be written in the form of the following equation:

which is far too short to use for further explanations, as one must understand which of three main transitional forms takes part in the reaction:

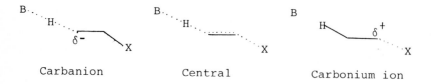

Carbanion   Central   Carbonium ion

Understanding is further hampered by an oversimplified reaction course:

$$d[E]/dt = k_1[KOH][PVC]$$

This equation suggests that there is only one essential constant of reaction rate, when we know that semi-products, denoted in differential equations by P_1, P_2,..., P_n, would have the following structures:

PVC $R-CH_2-CHCl-CH_2-CHCl-CH_2-CHCl-R_1$

P_1 $R-CH=CH-CH_2-CHCl-CH_2-CHCl-R_1$

P_2 $R-CH=CH-CH=CH-CH_2-CHCl-R_1$

.
.
.

P_n $R_2-(CH=CH)_n-R_3$

Therefore, each succeeding step of the reaction should be affected by a new charge distribution due to the growing number of conjugated double bonds in the direct neighborhood of the reaction site. Some years ago, Bunnett (64) formulated the simple principle of reactivity changes. He said that if any new bond was formed or broken causing new charge distribution due to the chemical change which would affect the activation process, it would also affect the transition state. Such conditions are evident when we compare, for example, the structures denoted by PVC and P_1. In the case of P_1 we have the electron-withdrawing group in close contact with the reaction site; therefore, the reaction cannot proceed in the same manner as in the first step. One might also find confirmation of this assumption in work by Östensson and Flodin (65), who observed that PVC degraded chemically is very vulnerable to further dehydrochlorination, proceeding even at 323K, which never occurs in the case of initial polymer. According to Ostensson (65), if dehydrochlorinated samples are stored for more than one year at room temperature, they eliminate HCl and turn darker. Furthermore they only swell in THF but do not dissolve. Definitely more than one constant of reaction must exist, and more than one mechanism of reaction must be expected in the course of reaction.

Further discussion of the possible reaction mechanisms should begin with the assumption that E2 mechanism does not follow a completely uniform pattern. We should rather think that all three transition states are in a kind of balance, and the gradually changing conditions of reaction are able to influence that balance. There is not enough experimental evidence yet presented to delineate the mechanism based on existing data; therefore, one can only use analogy for reacting components of low molecular weight.

From studies quoted above (31, 56) one knows that reaction starts from particular, not yet established, "weak points" in the chain because we finally arrive at a block-copolymer-like structure. If we assume that such a "weak point" exists in the case of bonds with a tertiary carbon atom , the elimination reaction might be initiated according to the E1 mechanism with formation of an intermediate carbonium ion, which is typical for tertiary alkyl halides. Then the leaving group (HCl) will stabilize the reactant and carbanion structure, and an increase in C-H bond breaking will follow. We might conclude, therefore, that in the second step, and probably in further steps as well, the reaction will be due to the E2 mechanism with carbanion as an intermediate. Additional arguments implicate that mechanism. The groups of electron-withdrawing character at the β-carbon atom to reaction site should increase carbanion character; the same trend parallels the base strength increase. Based on these assumptions, we might write the probable mechanism as follows:

1st step:

$$R_2-H_2C-HC-H_2C-\underset{\underset{Cl}{|}}{\overset{\overset{R_1}{|}}{C}}-CH_2-CHCl-R_3 + B$$

$$[R_2\text{-}H_2C\text{-}CH\text{-}CH\overset{\overset{\text{H}}{|}}{\underset{\underset{\text{Cl}}{|}}{C}}\overset{\overset{R_1}{|}}{\underset{\underset{\overset{..}{\text{Cl}}}{\delta^+}}{C}}\text{-}CH_2\text{-}CHCl\text{-}R_3]B$$

$$\downarrow$$

$$R_2\text{-}H_2C\text{-}CH\text{-}CH\overset{\overset{\text{H}}{|}}{\underset{\underset{\text{Cl}}{|}}{C}}\overset{\overset{R_1}{|}}{\underset{+}{C}}\text{-}CH_2\text{-}CHCl\text{-}R_3 + Cl^- + B^+ + A^-$$

$$\downarrow$$

$$R_2\text{-}H_2C\text{-}CH\text{-}CH=CH\overset{\overset{R_1}{|}}{\underset{}{}}\text{-}CH_2\text{-}CHCl\text{-}R_3 + HCl + BA$$
$$\underset{\text{Cl}}{|}$$

2nd step:

$$R_2\text{-}H_2C\text{-}CH\text{-}CH=C\overset{\overset{R_1}{|}}{\underset{}{}}\text{-}CH\text{-}CH\text{-}R_3 + B$$

$$[R_2\text{-}H_2C\text{-}CH\text{-}CH=C\overset{\overset{R_1}{|}}{\underset{}{}}\text{-}\overset{\delta^-}{CH}\text{-}CH\text{-}R_3]B$$

$$\downarrow$$

$$R_2\text{-}H_2C\text{-}CH\text{-}CH=C\overset{\overset{R_1}{|}}{\underset{-}{}}\text{-}CH\text{-}CH\text{-}R_3 + BH^+$$

$$\downarrow$$

$$R_2\text{-}CH_2\text{-}CHCl\text{-}CH=CH\text{-}CH=CH\text{-}R_3\overset{\overset{R_1}{|}}{\underset{}{}} + HCl + BA$$

The reaction might follow the mechanism of the second step until the termination step, as was discussed above. The reasons for termination of the polyene are similar to those discussed in the case of thermal degradation.

3.4. Comparison between various modes of PVC degradation

Chapters Two and Three contain data on PVC degradation by thermal energy of lower and higher levels, photolysis, irradiation and chemical elimination of HCl. The following summary is intended to delineate some differences and similarities. There is basically only one similarity between all modes of PVC degradation: polyene propagation. The effect of charge distribution on the elimination reaction is so strong that it applies to every kind of energy supplied, regardless of its level. Even combustion and pyrolysis seem to be a continuation of changes occurring in degradation below 473K.

Three groups of processes occur in the initiation reaction. Chemical elimination appears to be independent of the initial polymer chemical structure and more closely related to the total composition of the reacting mixture. Labile structures in the initial polymer are reflected in the degradation rate in the case of PVC thermal degradation in the lower temperature range. Photolysis and thermooxidation use the same labile points to start dehydrochlorination, but later they develop their own dehydrochlorination centers due to radical processes and the presence of oxygen-containing groups. If one were to judge the last two processes from a comparison between pyrolysis and combustion, one would conclude that oxygen involvement is related to an increase in the dehydrochlorination rate. This is true to a certain extent, but since in both thermooxidation and photolysis more time is allowed for reactions, their products can retain chemical structures of a more complicated nature due to the process involving oxygen and crosslinking, as in the case of photolytic reactions.

All these processes always affect each other if they are applied in sequence. For example, the thermal degradation rate is increased when the process follows either chemical elimination or irradiation as in Fig. 3.28. The increase in degradation rate is due to the oxidation process, which leads to increased radical concentration, as can be seen from Fig. 3.29.

These observations are not so important for polymer durability under practical conditions of its use since such a sequence can only appear when needed (e.g., in order to accelerate decomposition of wastes). The most frequent sequence in practice includes the photolytic process, following thermooxidation, which is probably the worst combination, since thermooxidative changes are always related to polyenes and oxygen-containing groups, and both are active in UV light absorption. Fig. 3.30 shows the effect of triplet and singlet oxygen on polyene decomposition.

The so-called oxygen photobleaching effect is characterized by data shown in Fig. 3.31.

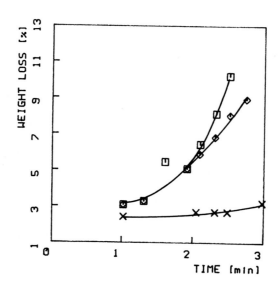

Fig.3.28. Irradiated PVC thermal degradation at 423K. × - control (degr. in O_2), ◇ - 52.2 Mrad (N_2), □ - 52.2 Mrad (O_2). (Modified from Ref. 38.)

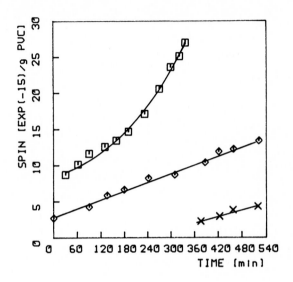

Fig.3.29. The effect of UV preradiation on spin generation for PVC during thermal dehydrochlorination at 473K (N_2). × - control, ◇ - irradiated for 1 hr, □ - irradiated for 3 hrs. (Modified from Ref. 28.)

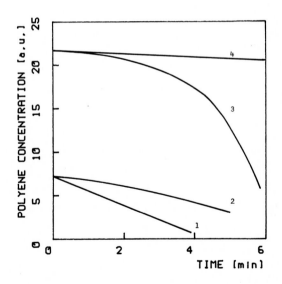

Fig.3.30. Decomposition of polyenes of different length due to oxidation by triplet (2) (3) and singlet (1) (4) oxygen. (3) (4) - 4 double bonds, (1) (2) - 10 double bonds. (Modified from Ref. 26.)

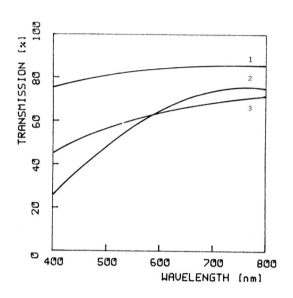

Fig.3.31. Oxygen bleaching of degraded PVC. 1 - control, 2 - 6 hrs exposure to UV in N_2, 3 - 8 hrs exposure to UV in moist oxygen. (Modified from Ref. 21.)

Although these processes help to recover initial color, at the same time the sample acquires so many oxygen-containing groups that weathering is enhanced. This shows how important the technological regime is for product durability. It is unfortunate that stabilizers have little effect on the elimination of UV-absorbing groups, with the result that costly additives must be used in order to achieve photolytic stability.

Because there are few connections between degradative processes, it is difficult to apply knowledge from one degradation mode to explain observations made for another. The same is true of results obtained in working with low-molecular models. Such results are frequently interpreted as if they were done for PVC, whereas, in fact, they are usually not related and merely confusing.

In future research we should probably treat every type of chemical change as a unitary process - a fragment of the overall mechanism of degradation. In looking at possible interconnections, we should try to prepare data applicable for mathematical modelling, since this is the only way to achieve an understanding of the phenomenon. To date this has not been done because most data reflect a functional relationship between two parameters, which obviously cannot give any answer for such complicated system.

REFERENCES

1. E.B. Mano and E.E.C. Monteiro, **J. Polym. Sci., Polym. Letters Ed.**, 19(1981)155.

2. W.H. Gibb and J.R. MacCallum, **Eur. Polym. J.**, 9(1974)771.

3. J.F. Rabek, Y.J. Shur and B. Rånby, **J. Polym. Sci., A-1**, 13(1975)1289.

4. J.F. Rabek, T.A. Skowroński and B.Rånby, **Polymer**, 21(1980)226.

5. V. Bellenger, J. Verdu and L.B. Carette, **J. Macromol. Sci.-Chem.**, A17(1982)1148.

6. V. Bellenger, J. Verdu, L.B. Carette, Z. Vymazalová and Z. Vymazal, **Polym. Deg. Stab.**, 4(1982)303.

7. D. Braun and S. Kull, **Angew. Makromol. Chem.**, 86(1980)171.

8. D. Braun and S. Kull, **Angew. Makromol. Chem.**, 87(1980)165.

9. F. Mori, M. Koyama and Y. Oki, **Angew. Makromol. Chem.**, 75(1979)123.

10. A. Torikai, H. Tsuruta and K. Fueki, **Polym. Photochem.**, 2(1982)227.

11. R. Salovey, R.B. Albarino and J.P. Luongo, **Macromolecules**, 3(1970)314.

12. R. Rånby, J.F. Rabek and G. Canbäck, **J. Macromol. Sci.-Chem.**, A12(1978)587.

13. F. Mori, M. Koyama and Y. Oki, **Angew. Makromol. Chem.**, 68(1978)137.

14. J. Verdu, **J. Macromol. Sci.-Chem.**, A12(1978)551.

15. S. Matsumoto, H. Ohshima and Y. Hasuda, **J. Polym. Sci., Polym. Chem. Ed.**, 22(1984)869.

16. B.B. Cooray and G. Scott, **Polym. Deg. Stab.**, 3(1980-81)127.

17. H.J. Heller, **Eur. Polym. J.-Suppl.**, (1969)105.

18. S. Hirayama, R.J. Foster, J.M. Mellor, P.H. Whitling, K.R. Grant and D. Phillips, **Eur. Polym. J.**, 14(1978)679.

19. E.D. Owen and I. Pasha, **J. Appl. Polym. Sci.**, 25(1980)2417.

20. R.J. Foster, P.H. Whitling, J.M. Mellor and D. Phillips, **J. Appl. Polym. Sci.**, 22(1978)1129.

164

21. J.W. Summers and E.B. Rabinovitch, **J. Vinyl Technol.**, 5(1983)91.

22. P. Dunn, D. Oldfield and R.H. Stacewicz, **J. Appl. Polym. Sci.**, 14(1970)2107.

23. A. Harriman, B.W. Rockett and W.R. Poyner, **J. Chem. Soc. (Perkin II)**, (1974)485.

24. E.J. Lawton and J.S. Balwit, **J. Phys. Chem.**, 65(1961)815.

25. N.-L. Yang, J. Liutkus and H.Haubenstock, **ACS Symp. Ser.**, 142(1980)35.

26. J.F. Rabek, B. Rånby, B. Östensson and P. Flodin, **J. Appl. Polym. Sci.**, 24(1979)2407.

27. E.D. Owen, I. Pasha and F. Moayyedi, **J. Appl. Polym. Sci.**, 25(1980)2331.

28. V.P. Gupta and L.E.St. Pierre, **J. Polym. Sci.**, **Polym. Chem. Ed.**, 17(1979)931.

29. P. Kohn, C. Marechal and J. Verdu, **Anal. Chem.**, 51(1979)1000.

30. E.D. Owen and J.I. Williams, **J. Polym. Sci.**, **Polym. Chem. Ed.**, 12(1974)1933.

31. E.D. Owen and R.L. Read, **J. Polym. Sci.**, **Polym. Chem. Ed.**, 17(1979)2719.

32. H.W. Gibson and J.M. Pochan, **Macromolecules**, 15(1982)242.

33. R.E. Michel, **J. Polym. Sci.**, **A-2**, 10(1972)1841.

34. T. Nakagawa, H.B. Hopfenberg and V. Stannett, **J. Appl. Polym. Sci.**, 15(1971)747.

35. S.W. Shalaby, **J. Polym. Sci.**, **Macromol. Rev.**, 14(1979)419.

36. W.H. Starnes, R.L. Hartless, F.C. Schilling and F.A. Bowey, **ACS Polym. Preprints**, 18(1977)499.

37. T.N. Bowmer, S.Y. Ho., J.H. O´Donnell, G.S. Park and M. Saleem, **Eur. Polym. J.**, 18(1982)61.

38. J. Guillet in **Polymer and Ecological Problems**, Plenum Press, N.Y., 1973.

39. G. Akay, T. Tincer and E. Aydin, **Eur. Polym. J.**, 16(1980)597.

40. R. Salovey and R.C. Gebauer, **J. Polym. Sci.**, **A-1**, 10(1972)1533.

41. M.A. Cook and W.R. Walker, **Nature**, 198(1963)1163.

42. J. Koch and W.R. Eckert, **J. Polym. Sci.**, **Chem. Ed.**, 15(1977)2791.

43. G. Lerke, I. Lerke and W. Szymański, J. Appl. Polym. Sci., 28(1983)501.

44. G. Lerke, I. Lerke and W. Szymański, J. Appl. Polym. Sci., 28(1983)513.

45. G. Lerke, I. Lerke and W. Szymański, J. Appl. Polym. Sci., 28(1983)519.

46. C.S. Marvel, J.H. Sample and M.F. Roy, J. Am. Chem. Soc., 61(1939)3241.

47. C. Sadron, J. Parrad and J.P. Roth, Comp. Rend., 250(1960)2206.

48. E. Tsuschida, C.N. Shih, I. Shinohara and S. Kambara, J. Polym. Sci., A2(1964)3347.

49. Y. Shindo, B.E. Read and R.S. Stein, Makromol. Chem., 118(1968)272.

50. W.I. Bengough and I.K. Varma, Eur. Polym. J., 2(1968)61.

51. Y. Shindo and R.T. Hirai, Makromol. Chem., 155(1972)1.

52. Y. Shindo, J. Polym. Sci. Polym. Letters Ed., 10(1972)555.

53. E.P. Chang and R. Salovey, Polym. Eng. Sci., 15(1975)612.

54. U. Schwenk, Angew. Makromol. Chem., 47(1975)43.

55. B. Östensson and P. Flodin, J. Macromol. Sci.-Chem., A12(1978)249.

56. A. Wirsén and P. Flodin, J. Appl. Polym. Sci., 22(1978)3039.

57. E.D. Owen and R.L. Read, Eur. Polym. J., 55(1979)41.

58. E.D. Owen and I. Pasha, J. Polym. Sci., Polym. Letters Ed., 16(1978)429.

59. B. Östensson and P. Flodin, J. Polym. Sci., Polym. Letters Ed.,

60. A. Caraculacu, J. Macromol. Sci.-Chem., A12(1978)307.

61. B. Östensson and P. Flodin, J. Appl. Polym. Sci.,

62. A. Wiersén and P. Flodin, J. Appl. Polym. Sci., 23(1979)2005.

63. R. Schlimper, Plaste u. Kaut., 13(1966)196.

64. J.F. Bunnett, Angew. Chem., Intl. Ed., 1(1962)225.

65. B. Östensson, in Dehydrochlorination and modification of poly(vinyl chloride), Ph.D. Thesis, 1979.

CHAPTER 4
PVC THERMAL STABILIZATION

The present monograph is aimed at discussing the phenomenon of PVC thermal degradation. We have already tried to study it from various angles, e.g., from the point of view of organic chemistry mechanisms, quantum chemistry, the structural properties of the polymer chain, and by comparison with photolytic, irradiation and chemical degradation. In the present chapter we should like to have a close look at the thermal properties of polymer in the presence of protective additives. Here we are to consider the topic, not really in connection with stabilization technology, but rather in the context of how stabilization affects the mechanism of PVC degradation compared with unstabilized polymer. Therefore, not the character of stabilizer, but its function and principle of action will be the main basis of discussion.

4.1. The mechanism of stabilization

HCl acceptance is unquestionably the most important feature of any chemical substance used as a thermal stabilizer of poly(vinyl chloride). Although there is basically no need to confirm this fact, since the catalytic effect of HCl is commonly accepted, many efforts have been made in order to study this phenomenon. The following set of chemical reactions shows the extent of the process:

- metal soaps:

$$M(OOCR)_2 + HCl \longrightarrow ClMOOCR + RCOOH$$
$$ClMOOCR + HCl \longrightarrow MCl_2 + RCOOH$$

- organotins:

$$(1) \quad R_2Sn(OOCR')_2 + HCl \longrightarrow R_2Sn(Cl)OOCR' + R'COOH$$

$$R_2Sn(Cl)OOCR' + HCl \longrightarrow R_2SnCl_2 + R'COOH$$

(2)

$$R_2Sn-OOCCH=CHCOOH + HCl \longrightarrow R_2SnCl_2 + (CHCOOH)_2$$

with the Cl substituent on the Sn.

(3) $\quad R_2Sn(SR'')_2 + HCl \longrightarrow R_2Sn(Cl)SR'' + R''SH$

$\quad R_2Sn(Cl)SR'' + HCl \longrightarrow R_2SnCl_2 + R''SH$

where:

 R - aliphatic group (mainly: methyl, butyl and octyl),

 R'- carboxylic acid rest (mainly lauric or monoisooctyl maleate),

 R''-mainly isooctyl thioglycollate, isooctyl thiopropionate, thiododecane.

- phosphites:

$$(RO)_3P + HCl \longrightarrow (RO)_2PHO + RCl$$

$$(RO)_2PHO + HCl \longrightarrow ROP(OH)HO + RCl$$

$$ROP(OH)HO + HCl \longrightarrow H_3PO_3 + RCl$$

- epoxidized compounds:

$$-CH-CH- + HCl \longrightarrow -CH-CH-$$

The reaction schemes suggest that the reaction proceeds in stages, but the naturally-observed phenomenon seems to be of a rather more complicated nature. It is known from Michel's earlier works (1, 2) with white lead that the process of HCl reaction goes through intermediate stages. These studies, recently repeated by Ball (3) for lead carbonate with careful analysis of reaction products by X-ray diffraction, proved without any doubt that lead chloride was the only product of reaction with HCl which formed no intermediate.

Similar studies for metal soaps are still in progress, and suggestions so far presented are even more controversial. The question to be answered in this case is: What is the order in which the reactions proceed? Two options are possible. Either HCl reacts first (the more probable option) with the stabilizer in its initial form $[M(OOR)_2]$, producing, first $M(Cl)OOR$ (which would suggest that the initial form is more reactive towards HCl than partially converted product is). Or partially reacted stabilizer is so reactive that a second act of HCl acceptance immediately follows the first one. Many authors (4-10) observed from their studies that the first possibility is correct. This author (11) studied the kinetics of reaction of HCl with various metal stearates in alcoholic solutions, and kinetic data suggested two different reaction rates, which should normally be regarded as confirming the first option. The results were interpreted in an earlier work (12) according to the same assumption.

All the above suggestions were made, however, without the most essential proof, i.e., isolation of chloro-metal carboxylate. A recently published paper (13) tried to fill that gap and isolate controversial intermediates. Studies were done for both hydrolysis and synthesis of metal stearates of Ca, Ba, Zn and Cd in various media. In water no compound other than normal metal stearate was detected, and therefore, studies were repeated in alcoholic solution. Results, as confirmed by IR spectra and elementary analysis, were exactly the same. The conclusion, therefore, was that probably the polar character of solvent does not allow one to obtain the intermediate product, whose change in symmetry is a destabilizing factor in such a media. The studies were repeated in water-free acetone, and an intermediate found for Ba and Ca stearates but not for Cd and Zn.

To further confirm these observations, another experiment was planned in which PVC and calcium stearate samples were sealed in extreme ends of the same glass tube, and the side containing PVC inserted into an oven so as to form HCl, which can be reacted by the thermal stabilizer. Several such tubes, processed under the same conditions, were kept for varying times to achieve varying conversions of calcium stearate. The resulting calcium stearate was immediately analyzed by X-ray and later for the contents of

Cl, which were indirectly determined by analyzing the consumption of Ag ions by AAS. Results confirmed previously-obtained data, since 81.3% of the reacted calcium stearate did not contain a trace of this product, but when alcohol was added to the sample and evaporated again, adequate amounts of calcium stearate were recovered, which shows both the possibility of reaction between calcium stearate intermediate and its lack of stability in alcoholic solution, as was found in another experiment.

How do we interpret these observations? As a matter of fact, the results of one work not confirmed by other studies, are concluded here but on the other hand no one has ever proved also that intermediates really exist either; therefore, it is assumed for further discussion that the reactivity of Ca and Ba intermediates is lower in non-polar media than that of initial carboxylates, and thus they can be stable in reaction conditions, which is not the case with the intermediates of Zn and Cd, which are expected to react immediately with a second molecule of HCl.

Considering the participation of organotin stabilizers in reaction with HCl, one question is important. Does reaction lead finally to $SnCl_4$, or can the reaction be performed only with groups other than alkyl? This question has been considered by many authors, and the overwhelming majority believe that HCl does not react with the C-Sn bond (6,14-16). Only Hoang (17) found - for model compounds - that reaction of aryl-substituted organotins may result in the formation of $SnCl_4$; Rockett (16) had explained earlier that this reaction might occur only when C-Sn bonds are cleaved due to the presence of, for example, phenyl rings. One should notice that for products of type $RSnY_3$ or R_3SnCl, the final product of reaction would be $RSnCl_3$ or R_3SnCl, respectively. There is still little doubt of the correctness of this conclusion, since it is known that, for example, R_2SnCl_2 retards PVC dehydrochlorination (18), but as the phenomenon is explained by reaction other than with HCl, this mechanism will be dealt with later.

Epoxidized compounds are able to accept HCl, which is a fact confirmed by much research (12, 19-21). The case of organic phosphites is slightly more complicated, since it was discovered that only alkyl phosphites, due to the greater electron density at the center (22), can react with HCl (23, 24), while aryl phosphite can only form complexes with HCl (24).

From this discussion we can already observe the possibility that various forms of stabilizer reaction products with HCl may react among themselves. The studies (13) above seem to suggest that the following reaction took place:

$$2ClCaOOCR \longrightarrow Ca(OOCR)_2 + CaCl_2$$

when alcohol was added to the Ca stearate being reacted with HCl in 81.3%. Possibilities for a reaction of this type were cited in so many papers (7-9, 25-28) that there is a little doubt that this is a true factor of PVC stabilization. The peculiarity of this reaction is that the metal type also plays an essential role here, which causes the following reactions to take place:

$$ZnCl_2 + Ca(OOCR)_2 \longrightarrow Zn(OOCR)_2 + CaCl_2$$
$$CdCl_2 + Ba(OOCR)_2 \longrightarrow Cd(OOCR)_2 + BaCl_2$$

This phenomenon helps to recover zinc and cadmium carboxylates in their initial form. Organotins are also bound by the same principle; moreover, they tend to balance their chlorines, which is quite useful, as products substituted with more chlorines appear at the end of the process, and these are usually more toxic (29, 30). Based on the same principle, organic stabilizers may interact with other stabilizers´ reaction products with HCl if they are present in the same reaction environment. Phosphites are thought to be able to react with zinc chloride according to the scheme below (8, 23):

$$ZnCl_2 + 2P(OR)_3 \longrightarrow (RO)_2-P(O)-Zn-P(O)(OR)_2$$

Also, epoxidized compounds are able to interact in stabilizing compositions according to the following equation (20):

$$HCl + -CH-CH- \longrightarrow -CH-CH-$$
$$\qquad\qquad\quad \backslash O/ \qquad\qquad HO\;\;Cl$$

$$2\; -CH-CH- + M(OOR)_2 \longrightarrow 2\; -CH-CH- + MCl_2 + 2\; RCOOH$$
$$\quad HO\;\;Cl \qquad\qquad\qquad\qquad\qquad \backslash O/$$

Both reactions seem slightly different in result: Where the first reaction helps to bind metal chloride, the second portrays epoxy compound as a moiety participating in HCl transportation to the stabilizer.

We can see that, so far, we have not encountered too many problems in explaining the experimental observations, and this image of completeness, so far as stabilizer reaction with HCl is concerned, would remain if not for the fact that we should also expect to have data on the stabilizers´ reactivity with HCl. Although this subject belongs to the kinetics rather than to the mechanism of stabilizer´s action, for the mechanism it is also important to know the comparative data for various products which differ in chemical nature. Unfortunately, there is little knowledge on this subject. One can only find traces of information, which do not appear fully reliable.

It was concluded that at the end of the induction period, a mixture stabilized by Ba/Cd stabilizer cadmium stearate was reacted in 50%, while barium stearate reacted in 80% (31). According to another paper (32), polymer color is changed when 80-90% of Ca/Zn or Ba/Cd stabilizer reacts with HCl. The reactivity of organotins with HCl, in alcoholic solutions, was higher than that of metal soaps (11). When organotins were reacted with a model compound resembling PVC, their reactivity was very high (14, 33). Wirth (19) found that the absolute organic conversion was independent of stabilizer concentration, but Figge (34) observed that the dehydrochlorination rate was higher when stabilizer concentration had fallen below 0.5%. The reactivity of phosphites towards HCl was said to be lower than that of metal soaps (35).

No data are available on the reactivity of epoxidized compounds. Information on stabilizer reactivity with HCl is far too limited, and what is really difficult at the moment is that we do not know how to treat this problem analytically.

The possibility of reaction with HCl would not justify production of so many different stabilizers types, and also would not satisfy the real meaning of stabilization. By definition, these products should be able to increase the stability of polymer. There are actually two different types of stability calling for different remedies in each case; one is to decrease the dehydrochlorination rate, and the other is to prevent color changes from occurring. At best, we should be able to perform these two functions in one act since color change is due to the formation of polyenes and dehydrochlorination also leads to polyenes, but since we cannot completely stop dehydrochlorination, we have to take a practical approach and find means of repairing what dehydrochlorination does.

Let us first discuss the function of stabilizer, which is intended to decrease the dehydrochlorination rate. The reasons for PVC thermal instability were discussed in Chapter One, and it is reasonable to believe that stabilizers are meant to diminish their effect on polymer stability. The most probable offender causing PVC instability is the presence of β-chloroallyl groups, which can initiate a chain reaction of HCl elimination, leading to polyene length extention. Stabilizers are thought to be able to react with this β -chlorine atom. It should be noted from the beginning that there are two theoretical possibilities for β-chloroallylic group deactivation:

$$-CH=CH-CH- \xrightarrow{ROOH} -CH_2-CH-CH-$$

with Cl below the first group, and O, Cl below the product, with $R-C=O$ attached to O.

$$2 \ -CH=CH-CH- \xrightarrow{(ROO)_2 M} 2 \ -CH=CH-CH- \ +MCl_2$$

with Cl below the reactant, and O below the product with $R-C=O$ attached.

both protecting the formation of conjugated sequences. All logic to the contrary, there is no clear explanation as to which of these reactions is likely to take place. The most common view since Frye´s work (36-38) is that the second reaction occurs, which can be easily understood, taking into account the high reactivity of metal carboxylates with chlorine. This view is supported by many recent works (10, 24, 39). Similarly, organotins are able to replace chlorides (40):

$$2 \ -CH=CH-CH- \ + \ R_2SnY_2 \longrightarrow 2 \ -CH=CH-CH- \ + \ R_2SnCl_2$$

with Cl below the reactant and Y below the product.

regardless of the structure of the Y group.

Guyot (24) believes that the reaction of chlorine replacement by carboxylic acid rest is catalyzed by $ZnCl_2$, which forms an intermediate according to the following equation:

$$2\ CH_3-CH=CH-CH-CH_2-CH_3 + 2\ ZnCl_2$$
$$\underset{Cl}{|}\ \downarrow$$

$$2\ CH_3-\overset{\delta^+}{\underset{\overline{}}{CH}}-CH-CH-CH_2-CH_3\cdot\overset{\delta^-}{Zn}Cl_3$$

$(RCOO)_2M$

$$2\ CH_3-CH=CH-CH-CH_2-CH_3 \qquad\qquad 2\ CH_3-CH=CH-CH=CH-CH_3$$
$$\underset{\underset{O=C-R}{|}}{\overset{O}{|}} \qquad\qquad\qquad\qquad\qquad\qquad + 2\ HCl + 2\ ZnCl_2$$

$$+ 2\ ZnCl_2 + MCl_2$$

As seen from the reaction, results come from studies of a low-molecular model, and suggest two possibilities for further reaction of an intermediate. Guyot (24) stated that replacement goes on as long as stabilizer is present, and when it is exhausted, the reaction pattern changes to rapid elimination, which leads to abrupt deterioration of the sample. A similar mechanism is thought to apply to cadmium carboxylates, but this is not the case with calcium and barium carboxylates (24). From studies by Braun (9), the sequence of reactions would include the following stage:

$$-CH=CH-\overset{\delta^+}{CH}-CH_2- \longrightarrow -CH=CH-CH-CH_2-$$
$$\underset{Cl}{\overset{|}{\underset{\displaystyle\left(\overset{..}{\underset{..}{:}}\right)}{|}}}\ \delta^- \qquad\qquad \underset{R-C=O}{\overset{O}{|}}$$

$$RCOOMOOCR \qquad\qquad + ClMOOCR$$

From this author's studies referred to above (13), it appears that since the salt of formula ClMOOCR (when M = Zn or Cd) is so unstable that the above reaction should rather be seen as follows:

$$\overset{\delta^+}{-CH=CH-\underset{\underset{\overset{|}{Cl}\,\delta^-}{|}}{CH}-CH_2-} \longrightarrow -CH=CH-\underset{\underset{R-C=O}{\underset{|}{O}}}{CH}-CH_2-$$

$$ClMOOCR \qquad\qquad\qquad + MCl_2$$

This part of the mechanism would be regarded as confirmed, if not for the serious contradiction between two observations when an identical technique was used (i.e., IR spectroscopy), one by Wypych (13) and the other by Vymazal et al. (42, 43), who did find the presence of ClCdOOCR. This result is surprising in that no one ever produced evidence that his compound can be isolated, and Wypych tried to obtain it in various ways without ever succeeding. Also, considerations of molecular symmetry do not suggest that this compound can be stable.

On the contrary, the positive effect of Cl replacement by organotin rests has been proved without a doubt by Starnes (44), who demonstrated how PVC dehydrochlorination decreases while the amount of sulphur incorporated into the polymer chain increases. This partially confirms that sulphur linkages with carbon are more stable and not easily hydrolized by the HCl which is evolved, and also shows why organotins (especially sulphur-containing) are better stabilizers than metal soaps. This observation also sends as back to the mechanism of metal soap action. Minsker (45-47) repeatedly shows that incorporation of carboxylic acid rest and its removal is in a dynamic balance, unlike the process portrayed by Guyot (24), who believes that the direction of the reaction is changed due to stabilizer exhaustion. It appears that Minsker is absolutely correct, for if the second process were right, then $ZnCl_2$ should participate, as long as stabilizer is available, in the reaction of Cl replacement for carboxylic acid rest, which would guarantee a much longer induction period for mixtures stabilized by zinc carboxylates, as there should be no particular reason to change the rate of dehydrochlorination. We know from practice that PVC stabilized with zinc carboxylate is only stable for a short period; even then there are still many labile chlorines to be replaced. This suggests that both processes are working equivalently, and if this is so, there is always a certain number of labile structures produced which have to be reacted.

The possibility of reaction with double bonds, which have already been formed or are available in the PVC chain, was not clearly indicated in the literature in the case of metal carboxylates, in contrast to the character of action of organotins (42) and phosphites (23), which are believed to take part in the following reactions:

$$R_2Sn(SCH_2COOR)_2 + HCl \longrightarrow R_2Sn(Cl)SCH_2COOR + ROCOCH_2SH$$

$$ROCOCH_2SH + -CH_2-CH=CH- \longrightarrow ROCOCH_2S(CH_2)_3-$$

and

$$-\underset{\underset{O}{\|}}{C}-CH=CH- + P(OR)_3$$

$$\xrightarrow{-ROR} \quad \overset{\frown}{O}-\overset{\overset{\|}{O}}{P}OR$$

$$\xrightarrow{+HCl}$$

$$-\underset{\underset{O}{\|}}{C}-CH_2-\underset{\underset{O=P(OR)_2}{|}}{CH}- + RCl$$

The reaction of thiols with double bonds was confirmed by measuring the unsaturations in PVC during treatment time; similarly, the reaction of phosphites was confirmed. Sulphur-free organotins were not able to react the double bonds already present, but they did not allow formation of new degradation sites since the level of unsaturations was constant throughout thermal treatment.

Summarizing the above discussion, some metal soaps (Zn, Cd) are able to replace the β-chlorine atoms, as are organotins and phosphites. Furthermore, sulphur containing organotins and phosphites can react with double bonds already available in the material. Present results cannot exclude this possibility for metal soaps. An investigator who wants to comment on this phenomenon must be able to measure simultaneously the rate of unsaturation formation, the rate of dehydrochlorination and the rate of free stearic acid formation.

We have already touched on the catalytic effect of some products of stabilizer reaction with HCl on dehydrochlorination

kinetics, but more explanations remain. Figs. 4.1 and 4.2 show
how the presence of various chlorides affects the PVC
dehydrochlorination rate (18).

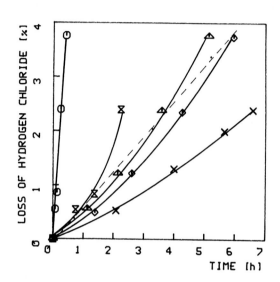

Fig.4.1. PVC thermal dehydrochlorination rate in the presence of
various chlorides. O - $ZnCl_2$; $\boxtimes$ - $CdCl_2$; Δ - $PbCl_2$; --- control;
$\Diamond$ - $BaCl_2$; $\times$ - Oct_2SnCl_2. (Modified from Ref. 18.)

The effect of the reaction of HCl with metal carboxylates is
typical, i.e., $ZnCl_2$ and $CdCl_2$ strongly affect
dehydrochlorination while the presence of other chlorides is
almost irrelevant. Except for the above explanation (24), there
is no more information on which we can rely to understand this
effect, and it is obvious that more studies are necessary to
enable a proper discussion of the catalyzed reaction mechanism.
In the case of organotin chloride, one should first mention that
another research group (48) obtained results which are not
exactly similar to those presented in Fig. 4.2, in the sense that
the presence of trichloride increases the dehydrochlorination
rate when compared with PVC alone. Wirth (18) explains the
effect of action of organotin chloride by its participation in an
allylic rearrangement:

$$-CH-CH_2-CH=CH-CH=CH-CH-CH_2-CH-$$
$$\quad | \qquad\qquad\qquad\qquad\qquad | \qquad |$$
$$\quad Cl \qquad\qquad\qquad\qquad\qquad Cl \qquad Cl$$

$$\downarrow$$

$$-CH-CH_2-CH=CH-CH-CH=CH-CH_2-CH-$$
$$\quad | \qquad\qquad\qquad\quad | \qquad\qquad\qquad |$$
$$\quad Cl \qquad\qquad\qquad Cl \qquad\qquad\qquad Cl$$

The newly formed structure is more stable, according to Wirth (18). Organotin chlorides are able to stabilize PVC color, which is attributable to the allylic rearrangement.

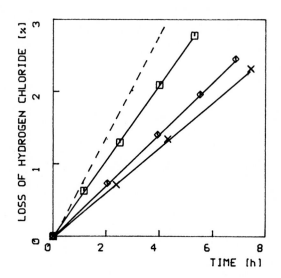

Fig.4.2. Thermal dehydrochlorination rate in the presence of various octyltin chlorides. -- - control; □ - $OctSnCl_3$; ◇ - Oct_2SnCl_2; × - Oct_3SnCl. (Modified from Ref. 18.)

These explanations would exhaust the participation of stabilizers if the PVC degradation process were based on a purely ionic mechanism, if it occurred in an inert atmosphere and if it included only dehydrochlorination. Since this is not the case, the discussion of stabilizer mechanisms continues in the next two paragraphs.

4.2. Oxidative and photolytic processes in stabilizers´ presence

In the Chapter Two, the difference between thermal degradation in inert and oxygen-containing atmospheres was clearly defined. Here, we shall merely examine the action of remedies and how they can help to protect the polymer. Nobody has commented so far on the participation of metal carboxylates in the process of polymer protection when oxidative changes occur. In the early years of PVC stabilization, antioxidants were added to the composition containing metal soaps until it was found that some of them might take part in further photolytic reactions. At present, the action of these stabilizers is certainly enhanced when they are used together with phosphites. Phosphites by themselves exhibit varying resistance towards changes of an oxidative character, as can be clearly seen from Fig. 4.3.

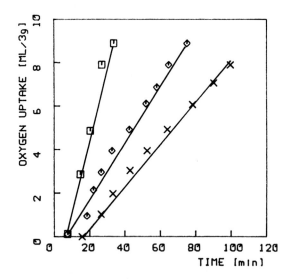

Fig.4.3. Self-oxidation stability of the phosphites. × - triphenyl phosphite, ◇ - diphenyl-isooctyl phosphite, □ - phenyl-diisooctyl phosphite. (Modified from Ref. 49.)

Phosphites reacting with hydroperoxides obey first order reaction-rate kinetics. In the presence of oxygen they retard the rate of PVC dehydrochlorination, which is probably due to the termination of thermooxidative reaction chains (23). Their reaction with peroxides is represented by the following equation:

$$R´OOH + P(OR)_3 \longrightarrow R´OH + O=P(OR)_3$$

Similar processes in the presence of organotins are more complicated, since there is a difference between sulphur-free and sulphur-containing organotins. Sulphur-free organotins are not able to react peroxides below a certain level; therefore, material stabilized with them always contains certain amounts of peroxides. On the contrary, sulphur-containing organotins can reduce the level of peroxide formation almost to nil, but not at the beginning of thermooxidation, when they even seem to increase the rate of peroxide formation. This initial increase in peroxide concentration is higher, depending on the amount of stabilizer introduced (41). Cooray (41) explains that the tin mercaptides convert hydroperoxides to tin sulphenates, which at elevated temperatures produce sulphur acids having an antioxidative action.

The other important function of stabilizers is that of radical deactivation. According to Mori (50), both metal soap and organotin stabilizers are effective in the deactivation of radicals, but organotins are more efficient than metal soaps. In the case of metal soaps, only those containing zinc and cadmium can participate in a deactivation reaction of a mechanism of an unknown nature. Ayrey and Poller (40, 51) presented a mechanism of radical deactivation which illustrates the participation of the alkyl group in this process. Their findings are interesting, since a similar mechanism describing the action of organotins during thermal degradation was proposed by Gupta and Pierre (52), who argued against the commonly approved view that the alkyl groups play an essential role in PVC stabilization due to reacting macroradicals formed during degradation.

The data on this subject are not yet numerous because the inclusion of oxygen complicates this phenomenon.

4.3. Polymer color

These few comments included here are meant to summarize previously discussed data. In some earlier studies it was noted with surprise that color change determinations and, for instance, dehydrochlorination rate, do not agree with each other. This phenomenon was partially explained by showing the mechanism of stabilizer action, which is capable of changing the length of polyenes; also thermooxidative processes, and especially photoinduced processes, may cause chemical and photolytic bleaching of polymer samples. Even if we include all these processes, we still should not expect that polymer color is the simple sum of all these effects. In order to realize this fact, we should refer to Chapter Two, where various side reactions are discussed. Some of these reactions are also efficient so far as color is concerned. If, for instance, benzene is formed in the degradation process, it means that the system has lost three double bonds, which otherwise would contribute to sample color.

As if this situation were not sufficiently complicated, in practical conditions we have two additional factors to consider that sometimes escape attention. One is connected with attempts to deal with color changes, regardless of chemical changes occurring by incorporation of dyes and masking agents, which have complementary color to the polyene's color. Iida (53) discusses this type of approach in detail. The other factor related to sample color is complicated in the more complex mixtures, usually including additives which may interact to cause color changes. It has already been discovered that such an interaction is possible between thermal stabilizer and UV absorbers (54, 55); also, it is known that benzyl-butyl phthalate may be converted to salol, which likewise contributes to color.

Finally, we should also underline the importance of conditions under which all degradation reactions are performed, i.e., the physical barrier of the chemical reaction which is occurring. This subject is discussed in more detail in Chapter Five, where the reasons for having so-called early color are attributed partially to the difficulties in performing necessary chemical reactions. It is also important in color stabilization to consider the dynamics of various interaction occurrences: stabilization process kinetics and the so-called synergism of stabilizer action.

4.4. Kinetics of stabilizers´ action

The kinetics of stabilizer action depend on its mechanism; in other words, on the functions a stabilizer is expected to perform. Genkina (56) discussed the kinetic relationships concerning the action of HCl acceptor in an enclosed volume. PVC dehydrochlorination in an enclosed volume is characterized by constant accumulation of HCl in the system, due to catalyzed decomposition. The HCl concentration change in polymer and gaseous phase is given by these equations:

$$\frac{\partial c_1}{\partial t} = D\nabla^2 c_1 + kc_1 + w_o$$

$$\frac{\partial c_2}{\partial t} = -D\frac{S}{V}\nabla c_{1S}$$

where:

c_1 - HCl concentration in the polymer,
c_2 - HCl concentration in gaseous phase,
D - HCl diffusion coefficient,
k - constant of catalytic reaction,
w_o - rate of uncatalyzed dehydrochlorination,
t - time,
S - specimen surface area,
V - volume of the system (sample + gas),
∇c_{1S} - HCl concentration gradient on polymer surface.

When the polymer is in powdered form, the above equations can be given as follows:

$$\frac{\partial y}{\partial t} = D\frac{\partial^2 y}{\partial r^2} + ky + rw_o$$

$$\frac{dc_2}{dt} = \frac{3Dg}{\rho VR^2}\left(\frac{y}{R} + \frac{\partial y}{\partial r}\right)_{r=R}$$

where:

  r - distance from grain center,

  $y = c_1 r$

$$S = \frac{4\pi R^2 g}{\rho 4/3 \pi R^3} = \frac{3g}{\rho R}$$

where:

  R - grain diameter,

  ρ - polymer density,

  g - aliquot.

In the beginning there is no HCl in the system; therefore, $y(r,0)=c_2(0)=0$; also, a condition on the polymer grain surface should be fulfilled such that $y(R,t)=Rc_2$ and $y(0,t)=0$ in the grain center (σ- "HCl solubility" in polymer).

For initial and limiting conditions, the concentration of HCl in a gaseous phase is given by the equation:

$$c_2 = \frac{w_o R^2}{\sigma D} \sum_1^\alpha \frac{\exp[k - (DZn^2/R^2]t}{Zn^2(1 + \frac{3\sigma g}{2\rho V}) + (\frac{kR^2}{D} - Zn^2)-(\frac{1}{2}\frac{kR^2/D - Zn^2}{3\sigma g/\rho V} - 1)} - \frac{w_o}{k\sigma}$$

where:

  Zn - equation root,

$$\frac{R^2 k}{D} - Z^2 = \frac{3\sigma g}{\rho V}(1 - ZcotZ)$$

Assuming that the HCl diffusion in polymer is very fast, we have the following relationship:

$$c_1 = \sigma c_2$$

When HCl acceptor is incorporated into the system, the total HCl concentration change is given by the equation:

$$dc/dt = w_o V_1 + V_1 kc_1 - V_1 k_1 [A]c_1$$

where:

V_1 - polymer volume,

[A] - HCl acceptor concentration,

k_1 - rate constant of reaction between the acceptor and HCl.

As total HCl concentration:

$$c = c_1 V_1 + c_2 V_2$$

and $c_1 = c/(V_1 + V_2/\sigma)$

where:

V_2 - gaseous phase volume.

therefore:

$$\frac{dc}{dt} = w_o V_1 - \frac{k_1[A] - k}{1 + V_2/\sigma V_1} c$$

As long as $[A] > k/k_1$, one can maintain stationary conditions for the degradation process in the presence of an HCl acceptor. This means that in stabilizer studies we are trying to determine the time at which a critical condition occurs:

$$[A]_{cr} = k/k_1$$

According to Genkina´s work (56) on the kinetics of lead stearate reaction with HCl concentration at initial concentration $[A]_o = 1.5 \times 10^{-1}$ mol/kg of PVC, the critical concentration (at induction period) is $[A]_{cr} = 2 \times 10^{-2}$ mol/kg, which means that 87% of stabilizer was used during the induction period.

Based on Minsker´s work (45-47, 57, 58), stabilization kinetics, in the presence of metal soaps able to substitute labile structures in PVC chains, can be explained by several kinetic equations. The rate of dehydrochlorination depends, in initial stages, on initiation (V_{in}) and propagation (V_p) rates:

$$V_{HCl} = V_{in} + V_p = k_{in} a_o + k_p \bar{\gamma}_o$$

where:

k_{in} – constant rate of random formation of β-chloroallyl groups,

k_p – constant rate of polyene propagation,

γ_o – concentration of labile structures (according to Minsker, β-carbonyl-allylic groups),

a_o – number of HCl molecules per one mer.

In the presence of agents having catalytic activity on the process (O_2, HCl, O_3 and so on), the equation should include their effect:

$$V_{HCl} = (V_{in1} + V_{p1}) + (V_{in3} + V_{p3})$$

where:

V_{in3} – rate of catalytic initiation,

V_{p3} – rate of catalytic propagation.

When metal soap is included in the mixture, it can substitute labile structures, but also, according to Minsker, the reaction in the opposite direction is in equilibrium; therefore, we can show the process kinetics as follows:

$$d[HCl]/dt = k_{in}a_o + k_p\bar{x}$$
$$dx/dt = k_h(\bar{\gamma}_o - \bar{x})$$

where:

$\bar{x}$ – current concentration of β-carbonylallylic groups,

$(\bar{\gamma}_o - \bar{x})$ – current concentration of substituted carbonylallylic groups,

k_h – rate constant of elimination of stabilizer rests and reformation of β-carbonylallylic group.

After integration and combining both equations, one can obtain the following relationship:

$$d[HCl]/dt = k_{in}a_o + k_p\bar{\gamma}_o(1 - \exp k_h t)$$

Minsker shows that k_h decreases with acid rest molecular weight increase ($CH_3COO-=0.35$; $C_3H_7COO-=0.21$; $C_{11}H_{23}COO-=0.12$ and

$C_{17}H_{35}COO-=0.12$).

The propagation of polyenes is described by the equations:

$$d[HCl]_n/dt = k_p\bar{x} + k_s\bar{x}d_o$$
$$d\bar{x}/dt = k_s\bar{x}d_o + k_h(\bar{\gamma}_o - x)$$

where:

k_s - rate constant for substitution of acid rest for chlorine adjacent to the double bond in the polyene being propagated,

d_o - initial stabilizer concentration.

We can see that in both initial and later conditions of dehydrochlorination, Minsker suggests the existence of dynamic equilibrium in stabilizer rest substitution and hydrolysis.

Combining the last two equations, one can arrive at the relationship:

$$V_p = \frac{k_p + k_sd_o}{k_h + k_sd_o}k_h\bar{\gamma}_o$$

The change in the number of chains due to crosslinking by MCl_2 effect is given by the equation:

$$dn/dt = V_{in} - V_c = V_{in} - k_c[MCl_2]\bar{\gamma}^{-\beta}$$

where:

k_c - crosslinking rate constant,

$[MCl_2]$ - concentration of metal chloride

$\bar{\gamma}^{-\beta}$ - concentration of isolated double bonds.

What is different in both kinetic models is that in the first, discussion concerned only the stabilizer playing the role of HCl acceptor; this means a stabilizer that should facilitate PVC thermal stability by diminishing the effect of HCl, but at the same time such a stabilizer should not produce metal salts having catalytic activity. In the second model, the stabilizer should be able to replace labile chlorine structures in the polymer, but at the same time it may produce metal chlorides having a catalytic activity affecting the dehydrochlorination rate and other processes (e.g. crosslinking). In both models, the catalytic effect of metal chlorides on dehydrochlorination was not included, which is correct for the first case, although it simplifies the real course of the degradation process. Minsker also did not include the HCl catalytic effect in his equations used for the analysis of stabilizer performance.

In his earlier works, Minsker (57, 58) used the following equation:

$$x = (k_1 + k_A d_o) a_o t$$

where:

k_1 - rate constant of uncatalyzed dehydrochlorination,

k_A - rate constant of PVC dehydrochlorination catalyzed by metal chloride,

in order to compare the effect of various metal chlorides on the dehydrochlorination reaction. Interesting conclusions were also reached by Minsker when he analyzed the catalytic effect of various products of stabilizer reaction with HCl, as can be seen from the discussion of stabilizers´ mechanism of action included above.

Still another approach to the kinetics of stabilizer action was presented by Wypych in a series of papers (11, 12, 28, 59-62). According to his model, a mixture of two stabilizers, e.g., one HCl acceptor and one able to substitute the labile structures in PVC chain, are analyzed together. The basic assumption, similar to the above equation, is that there is, for every formulation not including stabilizers, a characteristic rate of dehydrochlorination that is constant throughout the degradation period. Each stabilizer was assigned five characteristic rate constants, i.e.,

k_1 - reaction rate constant of $M(OOCR)_2$ with HCl,

k_2 - reaction rate constant of ClMOOCR with HCl,

k_3 - PVC dehydrochlorination rate constant under the influence of ClMOOCR,

k_4 - PVC dehydrochlorination rate constant under the influence of MCl_2,

k_5 - total cation exchange rate constant between initial, partially reacted and completely reacted stabilizers working together in mixture.

Based on this system of analytic and computational techniques, the values of the above constants were estimated and used for calculation of kinetic data of the stabilizers´ reactions in various systems. Recently, some of this model´s features were corrected as it was discovered that the mechanism of action of stabilizers is different than the one formerly assumed (13). Imputing the following experimental data: rate of uncatalyzed reaction, length of induction period, limiting capacity of stabilizer to react with HCl and calculated constants as above, one is able to compute the amount of HCl split off, the amounts of any stabilizer form present or re-reacting with other stabilizing components, and the amount of HCl emitted to the surroundings from the system versus degradation time. One can see from this model, that a set of kinetic data is available that broadly characterizes the changes occurring in the course of dehydrochlorination. Two simplifications are introduced to obtain a model which can be useful for computations, e.g., the fixed rate of uncatalyzed dehydrochlorination and the inclusion of the catalytic effect in the total value of this rate. These simplifications were necessary at the time of the model´s preparation and, unfortunately, still cannot be avoided, since there is no operative model of PVC dehydrochlorination. Still another problem hindering further development of mathematical modeling of PVC stabilization is the physical barrier of reaction, which will be discussed in the next chapter.

Since the above explanations mainly consider metal soaps, it is important to mention that Minsker´s kinetic equations were also applied to organotin stabilizers (63), and in principle, there is no difference in which type of stabilizer is analyzed by the above models; what is essential is the data to be included in the calculations.

Let us now analyze the outcome from modeling and other related studies which concern the kinetics of stabilizers action. Fig. 4.4 and 4.5 show the consumption of thermal stabilizer in the course of thermal degradation. From both graphs it is evident that there are at least two parameters determining stabilizer consumption, i.e., the chemical structure of the stabilizer and

188

its concentration. The effect of concentration is not a linear function of consumption; therefore, the assumption can be made that if the concentration of stabilizer is too low, it cannot efficiently bind the HCl that is able to participate in catalyzed dehydrochlorination, and that is probably why consumption of stabilizer is relatively higher.

The effect of the stabilizer´s chemical structure is explained by at least three factors: its chemical reactivity towards HCl, the catalytic activity of stabilizer or its reaction products with HCl so far as dehydrochlorination rate is concerned and the capacity for HCl acceptance of the stabilizer, related to the ratio of active elements. The presence of active stabilizing component has to protect the polymer against emission of HCl to the surrounding atmosphere, as can be seen from Fig. 4.6. Relative emission after the end of the initial period is lower due to the presence of unreacted stabilizer, which can also be seen from Fig. 4.4. The stability of polymer depends on stabilizer concentration, as shown in Fig. 4.7.

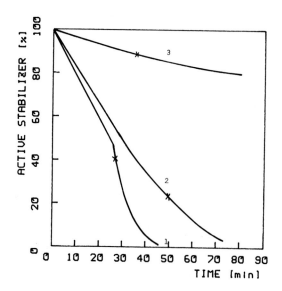

Fig.4.4. Stabilizer consumption in the course of thermal degradation. 1 - cadmium stearate, 2 - calcium stearate, 3 - tribasic lead sulphate. (x - end of induction period). (Modified from Ref. 12.)

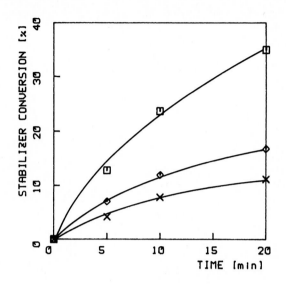

Fig.4.5. Dioctyltin bis(isooctyl thioglycollate) conversion versus degradation time at varying stabilizer concentrations. ▢ - 0.5%, ◇ - 1%, ✕ - 2%. (Modified from Ref. 19.)

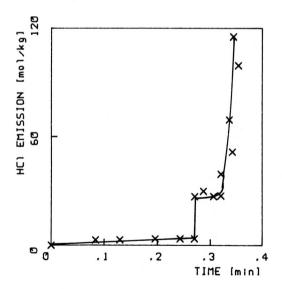

Fig.4.6. The kinetics of HCl accumulation during the degradation of PVC in enclosed space. PVC stabilized with lead stearate (0.15 mol/kg). (Modified from Ref. 56.)

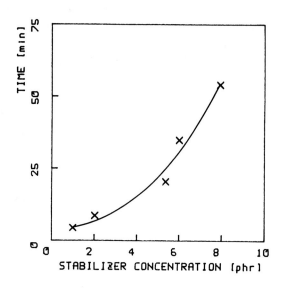

Fig. 4.7. The stability time as a function of the amount of tribasic lead sulphate. (Modified from Ref. 64.)

An interesting feature of these data is that increasing additions of stabilizer are gradually less effective in extending the induction period, which should be regarded as an effect of the accumulation of reaction products of stabilizer with HCl that are able to catalyze the dehydrochlorination. The accumulation of stabilizer products of reaction with HCl was determined recently (42, 65), as shown in Figs 4.8 and 4.9.

These two products of stabilizer reaction with HCl affect the dehydrochlorination rate, leading to increased consumption of stabilizer. It is interesting that calcium stearate used alone does not produce stearic acid, which has no other explanation than that the method used was not sufficiently sensitive to detect the relatively smaller amounts of stearic acid produced from stabilizer reaction with HCl.

The number of unsaturations produced is related to the concentration of stabilizer and its chemical structure, as shown in Fig. 4.10.

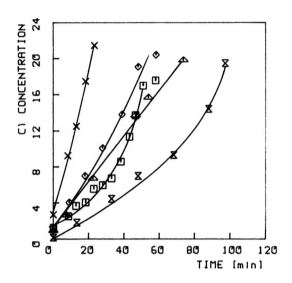

Fig.4.8. Chloride formation versus degradation time. × - Cd 2-ethylhexoate, ◇ - Ba 2-ethylhexoate, □ - Cd stearate, △ - Ba stearate, ⋈ - Ba/Cd stearates (1:1). (Modified from Ref. 42.)

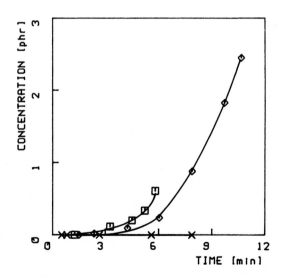

Fig.4.9. Concentration of stearic acid produced versus processing time. ◇ - Ca stearate 2.25 phr + Zn stearate 0.75 phr, □ - Ca stearate 1 phr + Zn stearate 2 phr, × - Ca stearate 3 phr. (Modified from Ref. 65.)

192

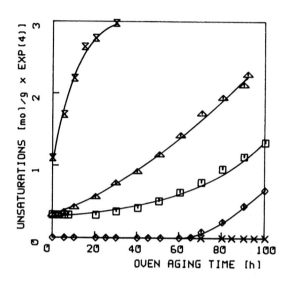

Fig.4.10. Formation of unsaturations versus degradation time. X - control, △ - dibutyltin maleate (5.8×10^{-3} mol/100g), □ - dibutyltin maleate (1.16×10^{-3} mol/100g), ◇ - dioctyltin bis(isooctyl thioglycollate) (5.8×10^{-3} mol/100g), × - dioctyltin bis(isooctyl thioglycollate) (1.16×10^{-2} mol/100g). (Modified from Ref. 41.)

Dioctyltin bis(isooctyl thioglycollate), which is able to substitute more efficiently the double bonds formed in the process, is more efficient in unsaturation deactivation. Also, the concentration of stabilizer has an important effect on reactivity with double bonds.

Fig. 4.11 shows the effect of the same stabilizers on peroxide concentration during thermal degradation. Peroxide formation, similar to unsaturation, is more effectively protected by dioctyltin bis(isooctyl thioglycollate). Also, in this case stabilizer concentration increase affects reaction probability and therefore decreases peroxide concentration (41).

Finally, we can also find data characterizing the kinetic effect of the action of several thermal stabilizers during photooxidation (Figs. 4.12 and 4.13).

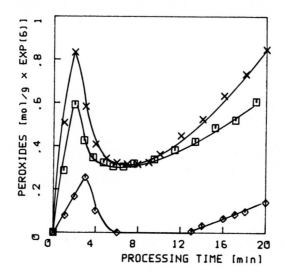

Fig.4.11. Effect of stabilizer type on peroxide concentration during thermal degradation. × - control, □ - dibutyltin maleate, ◇ - dioctyltin bis(isooctylthioglycollate). (Modified from Ref. 41.)

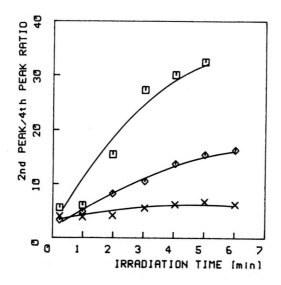

Fig.4.12. Change of ESR peak height ratio (2nd peak/4th peak) versus irradiation time. □ - Ba stearate, ◇ - Cd stearate, × - Zn stearate. (Modified from Ref. 50.)

194

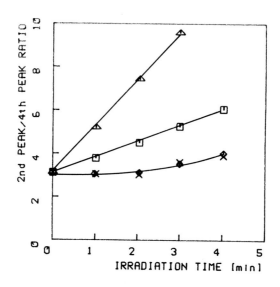

Fig.4.13. Change of ESR peak height ratio (2nd peak/4th peak) versus irradiation time.   Δ - control, ◇ - dibutyltin bislaurate, × - dibutyltin maleate,   □ - dibutyltin bisbenzyl-maleate. (Modified from Ref. 50.)

The ESR peak ratio characterizes the rate of oxidation in the sense that the lower the ratio, the lower the oxidation rate.   It is quite evident how big the difference is between both groups of stabilizers.   Both organotin stabilizers are known to be good for outdoor use, and they also protect the polymer well against oxidation.   Only zinc stearate, of the metal soaps studied, gives protection comparable to the polymer against oxidation.

Summarizing   the   above   data   concerning   the   kinetics   of stabilizers´ action, we can say that although the information is far from complete, we already have some achievements in both the modelling of properties and the results of experimental studies. The results would be more impressive if PVC thermal degradation without the presence of stabilizer were studied in more detail, especially if already-confirmed observations were quantified.

4.5. The nature of synergism

Synergism implies that two or more components, when used together, should perform better than is expected from their summarized actions when incorporated alone. This type of explanation may appear appealing when one wants to advertise industrial products and try to show the benefits of their application over similar products of the competition, but from the scientific point of view, to call diverse features by a common name creates a mystery surrounding simple, measurable observations which otherwise would be understood if taken one by one. Let us, therefore, point out the observations which have already been confirmed and explained and all other valuable suggestions which try to rationalize the meaning of synergism.

The first instance in such an explanation should perhaps concern various products of the reaction of stabilizer with HCl, since we can presume without any studies that they disturb polymer stabilization, as they act in the opposite direction to stabilizers. Metal soap stabilizers usually contain more than one chemical compound. The most typical metal compositions include Ca/Zn, Ba/Cd, Ba/Zn and Ba/Cd/Zn. If not for the catalytic effect of stabilizer reaction products with HCl, there would be no special reason to mix two different chemical compounds together, since cadmium and zinc carboxylates are more versatile than those of calcium and barium, as they can both accept HCl and react labile chlorine, thereby delaying color changes. The early color of PVC sample is improved when one of these two stabilizers (Cd or Zn soap) is increased, as shown in Figs. 4.14 and 4.15 for Ca/Zn and Ba/Cd stabilizers respectively. In these graphs, the lower the absolute value of the coefficient n, the better the color retention. At initial stages of heating, color is better preserved when a higher probability of reaction with labile chlorine exists, which means when a stabilizer of a lower ratio of Ca to Zn was introduced (total amount of stabilizer constant). When PVC samples are heated longer, the composition of stabilizer shows its adverse effect, as represented by Figs. 4.16 and 4.17 (24).

196

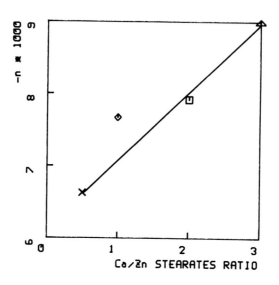

Fig.4.14. The effect of Ca/Zn stearate ratio on sample color.
The exponent of n of function $R=Ae^{nt}$ (R-reflection of light at
500 nm, A-constant, t-heating time at 448K) (66).

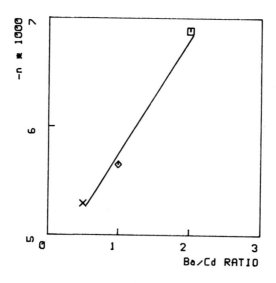

Fig.4.15. The effect of Ba/Cd stearate ratio on sample color.
The exponent n of function $R=Ae^{nt}$ (R-reflection of light at 500
nm, A-constant, t-heating time at 448K) (66).

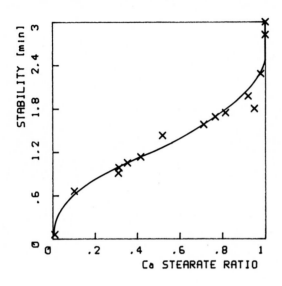

Fig.4.16. The effect of Ca/(Ca+Zn) ratio on action period. Degradation in Brabender plasticorder at 463K. (Modified from Ref. 24.)

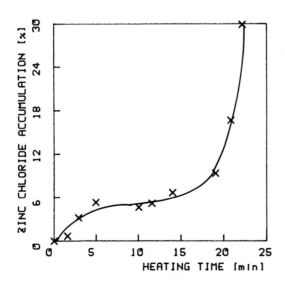

Fig.4.17. Zinc chloride accumulation in PVC during heating in Brabender plasticorder at 463K. Zn/Ca=1. (Modified from Ref. 24.)

The higher the fraction of zinc stearate, the shorter the induction period, since there is a better probability of $ZnCl_2$ catalytic influence on the dehydrochlorination rate.

From Fig. 4.17 we can see that zinc chloride concentration increases slowly at the beginning of thermal treatment and then remains at a more or less similar level up to the moment indicated on Fig. 4.16 as the end of its induction period. At this moment, the concentration of $ZnCl_2$ rapidly increases. Therefore, only one explanation is possible: Since degradation is in progress, the stabilizer reacts with the HCl, and at the same time, the color of the sample is preserved (see Fig. 4.14), which means that active degradation centers are inhibited and double-bond sequences not long enough to change the color. This explanation is summarized by the following reaction:

$$ZnCl_2 + CaSt_2 \longrightarrow ZnSt_2 + CaCl_2$$

and what is contained here is most important for a practical grasp of the principle of synergism. The reaction is correct from the point of view of basic chemistry. Similar observations concern Ba/Cd stabilizers as explained by Fig. 4.18.

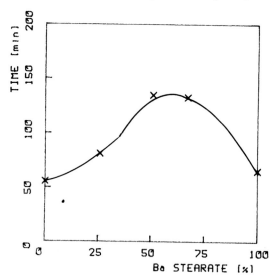

Fig.4.18. Induction period versus the percentage of Ba stearate in mixture with Cd stearate. Degradation temp. 473K. (Modified from Ref. 67.)

Practically all the investigators agree that a reaction between both stabilizing components takes place; therefore, it is surprising that Vymazal (5) obtained results that contradict this course of events. He quotes results of determination of total chlorine, cadmium and chlorine bound by cadmium which are compared with IR spectra of film obtained by sample dissolution in THF and evaporation of solvent. IR spectra show that in samples containing initially equimolar additions of Ba and Cd stearates, the amount of Ba stearate increases on heating (band at 1500 cm^{-1}) and that of Cd stearate (band at 1560 cm^{-1}) decreases. The first question is to how is it possible that Ba stearate, present in the sample from the beginning, does not show up on spectra, and only after 20 min of degradation does it start to appear slowly. We can discuss the other determinations them better when we examine the graph (Fig. 4.19).

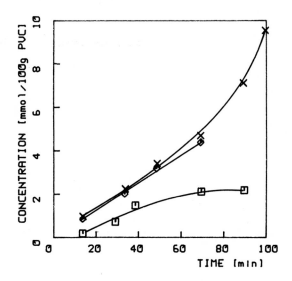

Fig.4.19. Total chlorides (×), chlorides in filtrate (◇), and cadmium (□) versus degradation time at 453K. Samples contain 5x10^{-3}mol each of Ba and Cd stearate/100g PVC. (Modified from Ref. 5.)

According to the method presented, the same authors (5) dissolved 1 g each of degraded sample in 80 ml of THF, added 2 ml of H$_2$0 (as explained, the water should facilitate dissolution of chlorides), filtered the solution, precipitated polymer with 170 ml of methanol and removed polymer by filtration. Filtrate, according to these authors, did not contain barium, which in any form existing in the mixture could not be soluble under the above conditions. This filtrate was used for determination of Cd and chlorine bound to it. It is not surprising, therefore, that the concentration of chlorine was twice as high as that of cadmium, since nothing else soluble in methanol can be formed but $CdCl_2$, as was explained in paragraph 4.1. There is another question not answered in the methods, i.e., how the authors determined total chlorine. More details about this paper were given here since it contradicts the commonly approved principle; therefore, one should analyze such a possibility more closely. It seems from the above analysis that, although it is difficult to say what was different in this study to cause their results to be incomparable with others available, the aforementioned doubts about their methods and results and should allow us to consider the commonly approved view as still being valid.

On this topic we should still refer to one of the most recent studies offering a new explanation of the action of Ca/Zn stabilizers. Mackenzie (65) suggests that these stabilizers act through a Zn stearate-Ca stearate complex. The possibility of forming mixed salts has been known for a long time. Structures such as:

$$[Zn(C_{18}H_{35}O_2)_4]^{2-}$$

$$[Zn(C_{18}H_{35}O_2)_6]^{4-}$$

$$[Cd(C_{18}H_{35}O_2)_6]^{4-}$$

can be found in many textbooks on inorganic chemistry. More recently, the possibility of formation of complex salts of Zn and Na was studied by Volka (68). According to Mackenzie's suggestion (65), polymer esterification occurs only if Zn stearate is not in complex form and before gelation. The first condition demands that Zn stearate is in excess of the quantity needed to form a complex salt, while according to the second condition after gelation, the mixture acts as an HCl acceptor only. Thus, the fate of the sample is decided before gelation occurs, while after gelation, the more zinc that remains, the worse for the sample's thermal stability. These conclusions do not quite agree with all the results already obtained, including some of those by the authors of the paper referred to, but at the same time they can better explain stabilizer efficiency when a mixture of metals is used.

These explanations do not exhaust the possibilities for deactivation of metal chlorides, which is a part of the umbrella mechanism called synergism. Fig. 4.20 explains the action of phosphites in this respect (69).

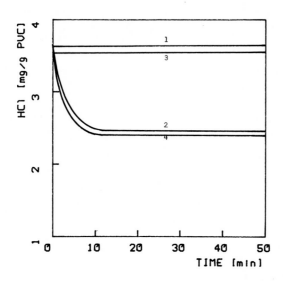

Fig.4.20. PVC dehydrochlorination rate at 448K. 1 - ZnCl$_2$, 2 - ZnCl$_2$ + tris(n-nonylphenyl)phosphite, 3 - CdCl$_2$, 4 - CdCl$_2$ + tris(n-nonylphenyl)phosphite. (Modified from Ref. 69.)

One can observe that in the presence of phosphite, both ZnCl$_2$ and CdCl$_2$ show less catalytic activity in PVC dehydrochlorination. According to Briggs (8), the action of phosphites is characterized by the equation:

$$ZnCl_2 + 2P(OR)_3 \longrightarrow (RO)_2P(O)-Zn-P(O)(OR)_2 + 2RCl$$

Based on Guyot's studies, this reaction is characteristic only for $ZnCl_2$ but not for $CdCl_2$ (24). Briggs (8) believes that polyols are also capable of binding zinc chloride, while Guyot (24) and Anderson (7), working with PVC model compounds, conclude that epoxy stabilizers are also able to bind $ZnCl_2$. Their point of view is not shared by other authors (21, 70), who rather suggest for epoxy compounds the role of HCl transportation and increased effectiveness of HCl reaction.

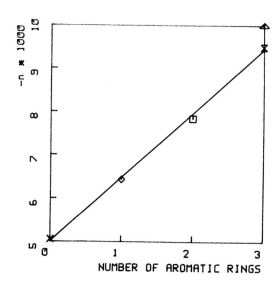

Fig.4.21. The exponent n of function $R=Ae^{nt}$ (R-reflection, t-degradation time at 448K) for PVC initial color change (PVC-100, $CaSt_2$-1, $ZnSt_2$-0.5, phosphite - 0.5). × - triisodecyl phosphite, ◇ - phenyl-diisodecyl phosphite, □ - diphenyl-isodecyl phosphite, ✕ - triphenyl phosphite, △ - tris(nonyl phenyl) phosphite (66).

The competition for reaction with HCl and labile chlorine is the other feature of synergism. For the sake of color retention we do not require that every act be due to reaction between the chain chlorine and stabilizer, especially when in this case too many groups of electron-withdrawing character would be introduced into the polymer chain. Thus, it seems that the number of acts of simple HCl acceptance balanced with those of substitution into the chain structures should produce an optimal solution. For this reason we also need to have a mixture of covalent and ionic carboxylates. The example of epoxy-compounds alters this balance towards substances able to react simply with HCl. Their role is even more complicated because some stabilizers cannot penetrate the grains, while epoxidized compounds do, which shortens the path of effective reaction with HCl. This phenomenon, as related to the physical barrier of reaction is discussed in more detail in Chapter Five. The reaction with HCl is also more efficient when metal soaps are used together with phosphites, as can be seen from Figs 4.21 and 4.22 (66).

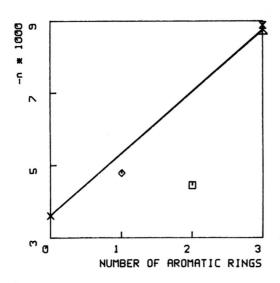

Fig.4.22. The exponent n of function $R=Ae^{nt}$ (R-reflection, t-time of degradation at 448K) for initial color change (PVC-100, Ba/Cd (1:2) stabilizer-1, phosphite-0.5). $\times$ - triisodecyl phosphite, $\diamond$ - phenyl-diisodecyl phosphite, $\square$ - diphenyl-isodecyl phosphite, X - triphenyl phosphite, $\triangle$ - tris(nonyl phenyl) phosphite (66).

For the initial color the most important considerations are to bind HCl effectively and to react labile chlorine structures, which are evidently enhanced by the presence of phosphites, since those able to react HCl (aliphatic) perform better.

There are many other examples of synergism that are not so well studied as the ones discussed so far. It was already mentioned when the mechanism of action of metal soap stabilizers was discussed, that they are probably not able to protect polymer against oxidation; therefore, when used with phosphites, their efficiency is increased since the presence of the carbonyl group would otherwise increase the dehydrochlorination rate.

Photolytic degradation is another example that is supported by many observations in practical applications that have not been scientifically interpreted. The presence of epoxy stabilizers always enhances the effect of action of metal soaps for outdoor use. Even the use of organotins together with epoxy stabilizers is more effective (71). This is explained by Hoang (71) by the catalytic action of organotin chlorides, due to which the epoxy compounds are able to replace allylic chlorine atoms. It is well-known that thioglycollates are not good stabilizers for outdoor use, whereas they are the most efficient thermal stabilizers. On the other hand, maleate esters are very efficient in outdoor stabilization. A mixture of the two can fulfill both demands. Burley (30) proposed a similar mechanism of chlorine exchange between initial and partially reacted organotins as is used for metal soaps. This mechanism is important for organotins since, on one hand, organotins, having three reactive groups are common synergistic additives, while, on the other, $RSnCl_3$ is more toxic and affects the dehydrochlorination rate more strongly; therefore, it is fortunate that $RSnCl_3$ appears at the end of the process.

Proper stabilizer choice can avoid difficulties in fulfilling other needs as well. For instance, in the case of barium-cadmium stabilizers, sulphur staining is a common disadvantage. By introducing a small amount of zinc carboxylate, staining does not affect the product, since zinc sulphide is white. This is another instance of synergism of action. The opposite phenomenon can be seen in the case of Ba/Zn stabilizers, which are usually very efficient, especially when initial color is concerned. But some mass and suspension polymers are so-called zinc-sensitive; incorporation of cadmium carboxylate usually solves the problem.

It seems that there are many examples of synergism which are either exploited in practice but not interpreted, or which may not even be noticed at all yet contribute to a certain product quality which is more efficient than others. This situation will be typical as long as more basic understanding is not available, and especially it will persist if both thermal degradation and stabilization are not expressed in quantitative form instead of descriptively, as at present. Recent studies offer some hope, since many basic problems have been resolved during the last few years, and it seems that research work in the field of PVC stabilization is soon to be more conclusive.

REFERENCES

1. E.W.J. Michel and D.G. Pearson, **J. Appl. Chem.**, 17(1967)171.

2. E.W.J. Michel, **J. Appl. Chem. Biotechnol.**, 23(1973)273.

3. M.C. Ball and M.J. Casson, **J. Appl. Chem. Biotechnol.**, 28(1978)765.

4. E.Czakó, Z. Vymazal and J. Štěpek, **Plast. Kaut.**, 9(1975)259.

5. Z. Vymazal, K. Volka, E. Czakó and J. Štěpek, **Eur. Polym. J.**, 17(1981)77.

6. P.P. Klemchuk, **Prog. Chem.**, 85(1968)1.

7. D.F. Andersson and D.A. McKenzie, **J. Polym. Sci.**, **A-1**, 8(1970)2905.

8. G. Briggs and N.F. Wood, **J. Appl. Polym. Sci.**, 15(1971)25.

9. D. Braun and D. Hepp, **Angew. Makromol. Chem.**, 44(1975)131.

10. A. Michel, **J. Macromol. Sci.-Chem.**, A12(1978)361.

11. J. Wypych, **J. Appl. Polym. Sci.**, 20(1976)553.

12. J. Wypych, **J. Appl. Polym. Sci.**, 23(1979)39.

13. J. Wypych, **J. Polym. Sci., Polym. Letters Ed.**, (1984).

14. G. Ayrey, R.C. Poller and I.H. Siddiqui, **J. Polym. Sci.**, **A-1**, 10(1972)725.

15. G. Ayrey, R.C. Poller and I.H. Siddiqui, **J. Polym. Sci.**, B8(1970)1.

16. B.W. Rockett, M. Hadlington and W.R. Poyner, **J. Appl. Polym. Sci.**, 18(1974)745.

17. T. van Hoang, A. Michel and A. Guyot, **Polym. Deg. Stab.**, 4(1982)213.

18. H.O. Wirth, H.A. Mueller and W. Wehner, **J. Vinyl Technol.**, 1(1979)51.

19. H.O. Wirth and H. Andreas, **Pure Appl. Chem.**, 49(1977)627.

20. J. Gilbert and J.R. Startin, **Eur. Polym. J.**, 16(1980)73.

21. J. Wypych, **J. Appl. Polym. Sci.**, 19(1975)3386.

22. T.B. Brill, **J. Organomet. Chem.**, 40(1972)373.

23. D.G. Pobedimskii, N.A. Mukmeneva and P.A. Kirpitchnikov, **Dev. Polym. Stab.**, 2(1980)125.

24. A. Guyot and A. Michel, **Dev. Polym. Stab.**, 2(1980)89.

206

25. P.P. Klemchuk, **Adv. Chem. Sci.**, 85(1968)1.

26. M. Onozuka, **J. Polym. Sci.**, A-1, 5(1967)2229.

27. R.D. Deanin, H.H. Reynolds and Y. Ozeayir, **J. Appl. Polym. Sci.**, 13(1969)1247.

28. K. Prochaska and J. Wypych, **J. Appl. Polym. Sci.**, 23(1979)2031.

29. R.E. Hutton and J.W. Burley, **J. Organometal. Chem.**, 105(1976)285.

30. J.W. Burley and R.E. Hutton, **Polym. Deg. Stab.**, 3(1980-81)285.

31. Z. Vymazal, E. Czakó, K. Volka and J. Štěpek, **Dev. Polym. Deg.**, 4(1982)71.

32. R. Nagatomi and Y. Saeki, **Japan Plast. Age**, 5(1967)51.

33. T. Suzuki, I. Takakura and M. Yoda, **Eur. Polym. J.**, 7(1971)1105.

34. K. Figge and W. Findeis, **Angew. Makromol. Chem.**, 47(1975)141.

35. F.J. Hybart and G.N. Rowley, **J. Appl. Polym. Sci.**, 16(1972)715.

36. A.H. Frye and R.W. Horst, **J. Polym. Sci.**, 40(1959)419.

37. A.H. Frye and R.W. Horst, **J. Polym. Sci.**, 45(1960)1.

38. A.H. Frye, R.W. Horst and M.A. Paliobagis, **J. Polym. Sci.**, A2(1964)1765.

39. D. Braun, **Pure Appl. Chem.**, 53(1981)549.

40. G. Aurey and R.C. Poller, **Dev. Polym. Stab.**, 2(1980)1.

41. B.B. Cooray and G. Scott, **Dev. Polym. Stab.**, 2(1980)53.

42. E. Czakó, Z. Vymazal, Z. Vymazalová, J. Skalsky and J. Štěpek, **Eur. Polym. J.**, 14(1978)1011.

43. K. Volka, Z. Vymazal, T. Zajiček and Z. Vymazalová, **Eur. Polym. J.**, 17(1981)1189.

44. H.W. Starnes, **Dev. Polym. Deg.**, 3(1981)135.

45. K.S. Minsker, S.W. Kolesov and G.I. Zaikov, **Vyssokomol. Soed.**, 23(1981)498.

46. K.S. Minsker, V.V. Lisitsky and S.W. Kolesov, **J. Macromol. Sci.-Rev. Macromol. Chem.**, C20(1981)243.

47. K.S. Minsker, C.W. Kolesov, E. Czako, A.P. Savelev, Z. Vymazal and E.M. Kiseleva, **Vyssokomol. Soed.**, 21(1979)191.

48. T. van Hoang, A. Michel and A. Guyot, **Polym. Deg. Stab.**, 3(1980-81)137.

49. E. Kovács and Z. Wolkóber, **J. Polym. Sci., Symp.,** 40(1973)73.

50. F. Mori, M. Koyama and Y. Oki, **Angew. Makromol. Chem.,** 75(1979)123.

51. G. Ayrey, F.P. Man and R.C. Poller, **J. Organomet. Chem.,** 173(1979)171.

52. V.P. Gupta and L.E.St. Pierre, **J. Polym. Sci., Polym. Chem. Ed.,** 18(1980)1483.

53. T. Iida and K. Gotō, **J. Macromol. Sci.-Chem.,** A12(1978)389.

54. J. Wypych, **J. Appl. Polym. Sci.,** 20(1976)279.

55. J. Wypych and Z. Pokorski, **J. Appl. Polym. Sci.,** 26(1981)1735.

56. L.G. Genkina and W.S. Pudov, **Vyssokomol. Soed.,** 24(1982)1919.

57. K.S. Minsker, W.P. Malinskaya and W.W. Sayapina, **Vyssokomol. Soed.,** 14(1972)560.

58. K.S. Minsker and W.P. Malinskaya, **Plast. Massy,** 3(1972)42.

59. J. Wypych, **Chem. Anal.,** 20(1975)233.

60. J. Wypych, **Polimery,** 20(1975)299.

61. J. Wypych, **Angew. Makromol. Chem.,** 48(1975)1.

62. K. Prochaska and J. Wypych, **J. Appl. Polym. Sci.,** 21(1977)2113.

63. K.S. Minsker, S.W. Kolesov and L.M. Kocenko, **Vyssokomol. Soed.,** 22(1980)2253.

64. W. Dick and C. Westerberg, **J. Macromol. Sci.-Chem.,** A12(1978)455.

65. M.W. Mackenzie, H.A. Willis, R.C. Owen and A. Michel, **Eur. Polym. J.,** 19(1983)511.

66. J. Wypych, **Proc. ACS Div. Polym. Mat.,** 52(1985)545

67. T. Nagy, T. Kelen, B. Turcsányi and F. Tüdös, **Polym. Bull.,** 2(1980)749.

68. K. Volka, Z. Vymazal, J.Stavek and V. Seidl, **Eur. Polym. J.,** 18(1982)219.

69. P.A. Kirpitchnikov, N.S. Kolyubakina and N.S. Mukmeneva, **Plast. Massy,** 7(1971)43.

70. J. Gilbert and J.R. Startin, **Eur. Polym. J.,** 16(1980)73.

71. T. van Hoang, A. Michel and A. Guyot, **Polym. Deg. Stab.,** 4(1982)427.

CHAPTER 5

THE CHEMISTRY OF STATE IN PVC DEGRADATION AND STABILIZATION

In earlier chapters we have discussed the degradation and stabilization of PVC from the point of view of purely chemical reaction without including the fact that all those reactions are performed in, let us say, peculiar conditions. In synthesis we tend to affect the reaction rate and yield by furnishing the reaction system with all possible means in order to make reaction possible, to enhance the reactivity of components and to achieve the preferable reaction equilibrium. The degradation processes are never planned this way as, at best, we would not like to see them at all; therefore, when they still dare to occur, we rather try to diminish their extent by lowering the thermal energy supply and combating all possible substances exhibiting some sort of catalytic activity rather then by affecting the reaction equilibrium by planned action. This is exactly the case in PVC thermal stabilization. As we know the possible reasons for thermal degradation and the reaction products, we try to bind them by other chemical reactions while less care is given to how it is done, which are the means to enhance the components reactivity and how to achieve satisfactory reaction equilibrium. This critical remark does not mean that we are absolutely careless so far as the last two groups of reaction parameters are concerned, but it underlines the character of studies available. In order to establish the accuracy of this statement, one should try to recall :

- how many publications have we seen on stabilizers´ reactivity with either HCl or groups in the PVC chain which are to be replaced;
- if there are papers discussing this reactivity in varying conditions (such as viscosity, temperature, amount of plasticizer and so on);

- how many studies consider the amount of stabilizer unreacted at the beginning of the initial period?

One may ask several more questions, but all will lead to the same conclusion, namely that the conditions under which stabilizers are chosen have been purely empirical ones till now; therefore, even if we consider these problems, we cannot get definite answers. From this author´s studies on initial color formation while stabilizing PVC with metal soaps and phosphites (1), it is estimated that over 70% of potential stabilizing capacity is never used due to the barriers of state in the stabilizers´ reactivity.

On the following pages we will try to analyze the possible barriers of state which can affect PVC degradation and stabilization. From the discussion to follow we can expect to know more about the conditions of HCl formation and catalytic activity from the point of view of structural barriers of material and of stabilizer-polymer, stabilizer-HCl and stabilizers´ reaction products-polymer inter-reactions also from the point of view of conditions of these reactions determined by material properties.

Discussion of the effect of the reaction environment on reactivity can conveniently begin with the influence of the viscosity of media on the chemical reaction rate since it relates to rheological studies. It would be interesting to consider some of Hildebrand´s controversial ideas (2) concerning viscosity and diffusion of gases in liquids. His theory differs from other theories concerning diffusion of gases in solids and liquids mainly because he does not agree that liquids have a lattice structure, which he contends was merely transferred without evidence from solids to liquids. According to Hildebrand,

> all molecules participate equally in thermal agitation that produces maximum disorder.

The particles diffuse simply because thermal motion keeps them ever on the move. Their mean displacement with time depends:

- upon temperature
- upon the ratio of intermolecular volume V, to volume, V_o, at which molecules become too closely crowded to permit either

diffusion or bulk flow.

The mean free path of molecules in simple liquids is very much shorter than their diameters; diffusion occurs by a succession of small displacements, which Hildebrand calls "random walk". Later it will be shown how molecules, especially polymer chains, can move. Now, considering that our reaction mixture contains various substances and that some of them are intentionally incompatible with others (lubricants), it is convenient for us to apply Hildebrand´s explanations since the transport of HCl through this mixture is described mainly by temperature and viscosity.

Shah (3) explains that melt viscosity is characterized by the following equation:

$$\eta = f(\gamma, \bar{M}_w, C, \mu, e, S, T)$$

where :

γ - shear rate,

c - plasticizer concentration,

μ - thermodynamic solvent power of plasticizer,

$\bar{M}_w$ - PVC molecular weight,

e - lubricant effectiveness factor,

S - stabilizer effectiveness factor,

T - temperature.

Let us discuss these factors from the perspective most useful for our problem. Shear rate may channel our explanations as different reactivity phenomena are observed on mixing and in free-path motion. The presence of plasticizer decreases viscosity of melt but, at the same time, may take part in HCl transportation and enhance stabilizer distribution. Maybe this is why plasticized PVC needs less stabilization, not only for the usual explanation that lower dehydrochlorination is probably due to dilution. On the other hand, the thermodynamic solvent power of the plasticizer would increase the viscosity of plastificate, and we can say that so-called fast-geling plasticizers have a detrimental effect on the PVC degradation rate. This effect can be explained by the more difficult diffusion of HCl and products of the stabilizers´ reaction which catalyze dehydrochlorination. The molecular weight of polymer affects not only the viscosity of melt or plastisol but also the flow pattern of melt. Pena (4) discovered that PVC melt has two different patterns of flow separated by a certain critical temperature. Below this temperature, PVC exhibits a rheological character typical of crosslinked polymer, while above it, PVC flows as does homogeneous melt. This observation is explained by the presence of crystallites below the critical temperature. It will be shown later that most probably the flow pattern is controlled by the grain structure.

A more detailed discussion of lubrication is necessary at this point. Lubricants are added to the polymer in order to reduce friction to a certain level. One type, an external lubricant, acts on the contacting surface between polymer and machinery walls. In order to perform its function, the external lubricant should not be compatible with polymer so as to migrate freely to its surface. The other type, called internal lubricant, supposedly lubricates, between polymer chains, which is not realistic since polymer grains are not even melted at this stage (see grain structure below); therefore, we can assume it lubricates polymer grains. We can view the whole mixture as containing polymer grains coated with lubricant on the surface,which if compatible, is found on the surface of macroparticles and penetrates into a complicated system of internal channels and cracks.

Now we come to the point frequently discussed in the literature. If we take one of many examples described in the patent literature (5), we shall see claims that thermal stabilizer becomes much more effective when lubricant is included than when stabilizer acts alone. In the case described (5), the stabilizer is four times more effective, which the authors attribute to the lowering of friction, and, at the same time, temperature decrease. Their explanation is probably correct, but why should one not also believe that three other effects are possible? First of all, the lubricant may be able to dissolve stabilizer, thus enhancing its distribution, allowing faster reaction with the HCl and diminishing the duration of its catalytic influence. Secondly, evenly distributed lubricant may isolate HCl and the eventual reaction products of the stabilizer from the polymer. Thirdly, because the lubricant is an

independently acting moiety on the border between two neighboring
PVC grains, it may facilitate HCl diffusion by channelling it
through the system, since it is liquid of lower viscosity than
that of the rest of the system. Therefore, if this "random
walking" of gas occurs, it would be more probable in the
direction in which less force is needed. Stabilizer effect, at
least as described by one author (3), acts on a principle similar
to lubricant. Finally, the temperature reduces viscosity, but it
is also able to increase it, as more polymer is either melted or
solvated. At higher temperatures, diffusion is faster, but the
amount of HCl produced and available is also higher; therefore,
the total amount of gas in the sample can even be increased.
Practical observation of how samples degrade under photolysis and
thermal treatment might explain the temperature involvement. It
is frequently observed that samples degraded under UV radiation
do not show uniform changes but usually have red or brown spots
of degradation. Similarly, when one compares thermally degraded
specimens including various stabilizers, one sees that those
containing zinc stabilizers also degrade in the form of spots,
which finally cover the entire surface area. Finally, some types
of suspension polymer are so-called zinc-sensitive. Perhaps all
these effects should be related to the diffusion or transport
phenomenon. In the first case it is lower temperature which
decreases diffusion, while in the last two cases it is the
presence of a very strong dehydrochlorination catalyzer ($ZnCl_2$),
which evidently cannot be effectively transported through the
system; hence, it affects certain areas more where degradation
started.

In order to relate the diffusion of gas, namely HCl, to sample
viscosity, we have just discussed factors connected with
viscosity, but there are also other factors affecting HCl
transport and other substances exhibiting catalytic activity.
One of them concerns the grain structure of PVC. In recent years
three excellent papers (6-8) have appeared on PVC grain
structure. All three, documented with photographs, lead us to
the same conclusion: that polymer, after being processed under
severe conditions of shearing at high temperature, retains most
of its original structure, especially when it has a wax and Ca
stearate coating, which can be seen from photographs to be

uniformly distributed. Suspension polymer has an especially complicated structure that includes submicroparticles (200-800A) agglomerated to microparticles (20-50 μm), and the last ones are encapsulated in a sort of sack forming the skin that holds other particles inside. Emulsion polymer, being sintered at elevated temperature, resembles large agglomerates having distinctive individual particles separated by channels. In order to destroy these structures at least partially, one investigator processed it for 18 min. at 463K in bicone, while all others heated polymer to 503K. Still, in both cases, the original structure could be recognized. The question is therefore, if we believe that these structures are not relevant to catalytic activities occurring on heating? Certainly we cannot neglect this parameter.

Let us now search the literature on PVC for possible effects of thermal treatment of a sample on its physical properties in order to find confirming arguments on the structure's participation in reactivity. The first information directly linking distribution of additives in PVC within the scope of our interest comes from Katchy's work (9). In his studies on dry-blend preparation, two different thermal stabilizers were included, one solid tribasic lead sulphate stabilizer, and one liquid organotin stabilizer. Dry-blends were prepared in a fast mixer with a controlled temperature that was allowed to rise, while mixing, up to 413K. Samples of dry-blends were analyzed by scanning electron microscopy among other methods. Dry-blend, including tribasic lead sulphate, showed that the stabilizer was found only on the grain surface without a trace inside the grain. The mass polymerizate showed that some stabilizer particles were able to penetrate larger internal channels but only approximately 2 μm from the grain surface. Similar studies of liquid tin stabilizer done by DTA measurement show that this stabilizer lowers T_g. According to Katchy's assessment (9), liquid stabilizer penetrates into the grain, but it is not evenly distributed.

Studies of a similar nature have been reported on partially soluble additives in PVC, e.g., stearic acid, butyl stearate and ethylstearate (10). In this research, a differential scanning calorimeter was used to determine the content of undissolved diluent. The samples were prepared either by dry-blending or calendering. The results confirm the very limited solubility of stearates with PVC, which increases with rising temperature. Many data are available on the distribution of solid additives in PVC melts from tests of extruder performance that are of high concern for producers of such equipment. From macroscopic observation it appears that good dispersion of additives is achieved, but magnified images also show absence of penetration into the polymer grain. The other, more drastic case, is recorded by X-ray studies of plasticized PVC. Many authors (11-15) have observed that plasticizer is unable to destroy crystalline organization. Furthermore, stabilizers cannot penetrate crystallites but are distributed in amorphous areas.

Recent studies include even more specific comments relating the dimension of a molecule to its reactivity:

> in case of carbonyl chromophores, triplet energy transfer between nearest vicinity in a macromolecules in solution occurs with rate constants in the range of $10^{10}-10^{12} s^{-1}$. Quenching of macromolecules by small molecules is usually slower than quenching process involving two small molecules, an effect that can be attributed to a decrease of the diffusional contribution of the energy donor (16).

Amer (17) describes the diffusion of HCl entrapped in polymer structure as a very slow process. The same author quotes the value of activation energy of krypton in PVC films as equal to 205 kJ/mol above T_g, which he believes to be the order of activation energy of HCl diffusion in PVC. The presence of HCl in the sample may alter the mechanisms of reaction occurring, for which examples can be found in many publications. For instance, Levai (18) relates increased HCl concentration to an increased rate of double-bond formation, while Owen (19) relates it to the formation of long sequences by accumulation of short ones due to migration of polyenes along the polymer chain.

The reaction rate between HCl gas and solid lead salts ($PbCO_3$ and $2PbCO_3 \cdot PbO \cdot H_2O$) is discussed by Ball (20). The constant rates of reaction are pressure dependent, which suggests that a diffusion process is the rate-controlling step. The reaction occurs on the surface, with HCl being a diffusing specie. The temperature increase is parallel to the rate constant increase, which is attributed to the change in diffusion coefficient. The other example of a diffusion controlled process is given in

Verdu's paper (21), which studied the effect of PVC film thickness on the rate of C=O group formation by UV irradiation. The rate increases, along with an increase in the polymer's film thickness, because the process is controlled by the diffusion rate of HCl out of the sample.

The ISO Technical Report (22) includes the graph shown in Fig.5.1.

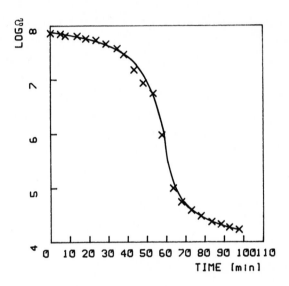

Fig.5.1. Electric conductivity of sample versus degradation time.

which comes from one of the commonly-used techniques to determine the stability of PVC samples, which is expressed by induction period. This method has been known for more than 20 years, and so far nobody has explained why the curve slopes down right from the beginning. Is it due to sample compression, formation of chlorides or gradual increase in the presence of HCl in the sample because of the increasing time interval and path length for the reaction between HCl and stabilizer? These questions cannot be answered at present because we lack sufficient data.

We can see that all the results quoted are derived as a side effect of work intended to explain other properties of PVC materials. At the moment one cannot find studies done with the intention of showing the effect of the reaction environment on thermodegradation and the stabilization process.

In further explanations we must consider theories of molecule
migration and diffusion in the viscoelastic state. The following
theory was formulated to describe the motion of polymer chains in
viscoelastic states, and it is applicable to polymer melts and to
the diffusion of polymers of various structures. Differences in
Brownian motion between the macromolecules, especially in
concentrated systems, and in low molecular weight substances
should be included. In the case of low molecular weight models,
motion is controlled mainly by distances between molecules, their
dimensions and their thermal state. In the case of
macromolecules, thermal energy plays the same essential role, but
it is difficult to talk about distances and dimensions the same
way as in the case of small molecules. Polymer chains have an
entangled structure, and when they move they should not pass each
other. In order to explain their motion, they are seen as chains
placed inside a tube-like region (23) (Fig.5.2).

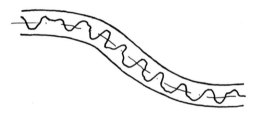

Fig.5.2. Chain segment.

The central line inside the "tube" is called the primitive path,
and it is always shorter than the real chain. If the ends of the
chain are fixed, for instance by crosslinkages, the deformation
of the polymer results only in deformation of the primitive
path. When chains are not fixed, as in most linear polymers,
diffusion is only allowed along the primitive path. In other
words, in a concentrated polymer system, each chain can move
independently in the field imposed by the other chains, and each
chain is confined in a tube-like region surrounding it. When the
system is at rest, the tube is fixed in space, the opposite of
macroscopical deformation wherein the tube is also deformed.
What is also important is that any change of polymer conformation
has to result in immediate change of the tube-like region. It
is, therefore, evident that the chain, in Brownian motion, moves

randomly forward or backward inside the tube relative to other surrounding macromolecules. This diffusion process depends on concentration and molecular weight, which means it is also related to the same factors as diffusion and Brownian motion of small molecules, but the route is more restricted.

It seems useful to present another model of network, the slip-link model (24). This model describes molecular motion under flow and should better explain polymer processing.

Fig.5.3. Slip-link model.

In this model we have small rings through which the chain has to pass. It is assumed that polymer moves freely in the space between the slip-links, and the slip-links might be regarded as points of more densely-packed structure through which the polymer chain must pass. The conformation change is easier when such a model is applied, and also it is easy to understand heat generation during flow.

The above models are not of purely theoretical importance since they agree with experimental studies of stress relaxation and - what is important for our discussion - with translational diffusion of polymer in melt (25). Gilmore (26) has measured the effect of temperature and molecular weight on macromolecular diffusion in blends of PVC and poly(β-caprolactone) by scanning electron microscopy and X-ray fluorescence analysis. The diffusion coefficient is inversely proportional to the first power of molecular weight. Temperature increase affects proportional increase in diffusion, and results obtained demonstrate that the activation energy of diffusion equals 49 kJ/mol. The diffusion coefficient was determined to be in the range 10^{-12}-10^{-13}cm^2/s. It is interesting to notice from studies on polyethene samples (25) that the topological environment of molecules in a melt (the "tube" surrounding them) only changes slowly during the diffusional motion. A recently-developed technique for measuring diffusion in polymer films by fluorescence redistribution after pattern photobleaching (27) raises hopes that in the near future we should have more data on diffusion in polymeric materials. Fluorescent dye is added to polymeric material and foil cast from solution. After an initial high-intensity burst of light (laser) in order to bleach the dye in the illuminated region, diffusion of fluorescent molecules into the bleached region is observed by measuring fluorescence intensity. Many computational techniques discussed in the literature can be used either for modeling or data computation (28-30).

Summarizing the above discussion on the possible diffusion or migration of molecules, we should observe first of all that these models resemble Hildebrand's thoughts on diffusion rather than the widely used theory of holes for solid materials. Secondly, one can certainly see that by introducing rigidity to the structure, either by crosslinks or the presence of rigid segments in a chain (e.g, all-trans polyenes), the system becomes less flexible and even micro or local deformations, more difficult. Thus, at more advanced stages of degradation, the structure will be increasingly "frozen", i.e., if there is any system of internal transporting channels, they will remain increasingly fixed. This may explain why at higher stages of degradation the rate is more uniform than at the beginning. The presence of small molecules such as plasticizers gives more freedom for "tube-finding", and that is why diffusion is far easier. Considering small molecule diffusion in the polymer system, we can describe it in a similar way to the above or by using the model for polymer-solvent interaction (31). Diffusion, in this case, depends on the molecular mass of both substances, their specific free-hole volume and the mass density of solvent. In this case, the size of the molecule diffusing in the polymer has a detrimental effect on the process; therefore, to take this effect alone, a metal soap of, let's say, lauric acid should work better than one of stearic acid.

The diffusion of gas, e.g., HCl, through polymer can be explained simply by imperfections of structure that are evident from all related studies such as scanning electron microscopy, X-ray structural analysis, density determination, BET isotherms and so on. Also, the above models contribute to the same belief: that PVC in the viscoelastic state has a certain fraction of free spaces. These imperfections are larger than the dimensions of the HCl molecule and it can, therefore, penetrate the structure by Brownian motion in order to balance concentration with the surroundings. In this case, the process would be rather slow (usually called "random walk") as the total path length of molecule per second is estimated to be about 1 m, but at the same time the effective distance covered by a particle through the sample thickness would be in the range of 10^{-5}m. The question immediately arises if this is a sufficient speed to release gas from the sample. We can answer this question by simple observation. In the beginning of the degradation process this rate and the reaction rate of HCl with stabilizer are totally sufficient; otherwise we would be unable to produce transparent foil. Later it is not sufficient, because the degraded sample always increases its volume, which was observed by all investigators who ever tried to degrade PVC material or powder.

Let us now leave this phenomenon for further discussion below, and follow the diffusion process. This process has still other complications, and one of them is adsorption on the surface of material. As Amer (17) has said, diffusion of krypton through PVC film above T_g is characterized by an activation energy of 205 kJ/mol, which is of the order of activation energy of the HCl diffusion process. If this is the case, it means that HCl diffusion is affected by chemisorption as activation energy is higher than 80 kJ/mol, which is usually regarded as the border-line between simple physical adsoption and chemisorption.

There is still an additional mechanism to be included so far as gas diffusion through the polymer is concerned. It is well known in solid state physics that when solid materials are heated below the melting point, grains are sintered in the process, and therefore diffusion gradually becomes more difficult. Moreover, in this process material-free enclosed spaces might be formed, which would definitely act as centers of degradation of increased rate. There are, of course, limits to structural compactness, especially when we also have gas emission due to the degradation process. Having a well-defined grain structure in the polymer under discussion, we should also remember that diffusion would not be uniform, as we would certainly expect to observe a path of high diffusion on the intergrain surfaces. It is known from studies on gas diffusion in metals that 25-50% of total diffusion is due to such a high diffusion path.

From this rather brief explanation we can see that the diffusion process of gas in solid material is highly complicated because of initial structure, chemisorption and dynamic changes in material structure related to the sintering process and local overpressures.

Let us return to the increased volume of the specimen during the thermal degradation process. In discussing this phenomenon, we would like to study closely the type of diffusion process that occurs from the point of view of the internal structure of the material that is being restructured by the internal process of gaseous product evolution. If we imagine the initial structure of the material as being formed by grains of close-to-spherical shape with spheres again constructed from agglomerates enclosed in envelopes, as is the real case in material processed from suspension PVC, we can see that there are two different material free spaces, such as those shown in Fig.5.4.

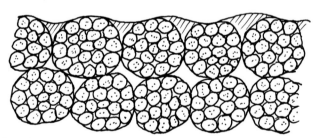

Fig.5.4. The structure of PVC material.

There are free spaces between the agglomerates, but these are enclosed in an envelope, and there are also free spaces between the grains or envelopes. The solid material contains crystalline structures, which have a high melting point (estimated to be 533K), and amorphous material, which has a lower melting point and dominates in polymer mass. When we start to apply heat to this material, after a certain period of time, the material should obtain a uniform temperature distribution; therefore, the degradation rate for conditions under inert gas should be stable, as either grains or microparticles are sufficiently large to be regarded as having equivalent distribution of all the internal defects which are to control the initiation of the degradation process. At the same time the stabilizer is either on the grain surface or eventually, at best, has penetrated the envelopes and is somehow distributed (not very evenly) inside the grain. Suppose we are to discuss the first case, that the stabilizer is distributed on the grain surface, which makes it more difficult for the reaction to take place but at the same time better explains the idea of such a model's action.

On heating, a gaseous product is formed equivalently throughout the mass of polymer as every particle is formed from the same statistical material. These HCl molecules formed on the surface are in the direct neighborhood of the stabilizer, and they will react immediately; perhaps there will be no gas formation, but reaction might proceed directly with substitution of stabilizer rests into the polymer chain. An opposite trend should be predicted for dehydrochlorination stages that are in the grain inside. This HCl has first to diffuse in the spaces between microparticles and agglomerates and then through the envelope; such a process is not immediate since diffusion is slow. If there is any stabilizer to react with, after the HCl molecule is released from the envelope inside, the HCl will be reacted; if not, it will increase its concentration in the space between the grains and tend to diffuse towards the surface of the material, which has the lowest concentration of this gaseous product. Based on this explanation, we can already say why there is no (or very little) HCl in the sample surroundings at the beginning of heating. The most privileged path for HCl diffusion at this stage is between the polymer grains, i.e., between the surfaces of grain envelopes, as there is no reason for HCl to diffuse inside those envelopes due to its higher concentration there. But this stage is not in favor of gas diffusion because we have a simultaneous process of sintering, since there are always molecules of lower melting-point trying to glue to the adhering surfaces. Therefore, the gas in the free spaces between envelopes increases in pressure, which affects the joint points trying to seal the structure, and finally, when the wall surrounding the HCl-filled space is thin enough, gas forces the wall and joins the next free space in its neighborhood. After several repetitions, gas is ready to diffuse outside the material. At first this process is of an "explosive" character, for in the meantime serious overpressure has been formed. This is why in the next stage of the degradation process we can monitor the sudden increase in HCl evolved as shown on Fig.5.5.

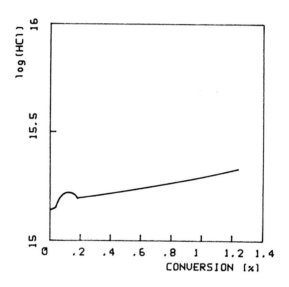

Fig.5.5. Dehydrochlorination stages.

Once we pass through this stage it does not mean that the process is stabilized; indeed, it is still erratic, since equilibrium is not yet achieved, but after a short while, the rate of HCl emission is stabilized due either to the formation of stabilized routes of gas escapement or to balanced channels closing and new ones opening. The studies made on the sample subjected to constant heating but, at the same time undergoing occasional changes in carrier gas flow rate, are referred to in Fig.5.6 (32).

After the rate of HCl emission has been stabilized in order to pass the stages mentioned in Fig.5.5 and a constant HCl emission rate has been achieved, the carrier gas flow rate was increased, resulting in a fast increase in HCl emission (second peak), which was gradually becoming constant when again the flow rate of carrier gas was increased and another sharp peak obtained. Every increase in flow rate resulted in such a peak, while when the

flow rate was decreased, the gas emission dropped down.
Increased flow rate of carrier gas facilitates the evacuation of
HCl from material inside. By contrast, a decreased flow rate of
carrier gas results in a delay before HCl concentration inside
the sample reaches equilibrium.

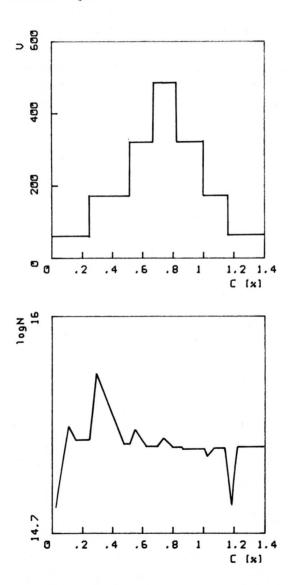

Fig.5.6. The rate of HCl emission (N) and the rate of carrier gas
flow (V) versus the conversion (C) (32).

It is worth noticing that a constant emission rate decreases on evacuation and increases on slowing down the carrier gas flow rate. It is difficult to say if this is due to the already-formed structure of channels or to a certain number of channels being statistically present throughout the process, but from the immediate reaction on flow rate change, we can suggest that channels and internal chambers filled with HCl are already present.

Based on present knowledge, one is not able to predict the real nature of channels, i.e., if they are stationary or if they appear and disappear. Their nature depends on several factors, but first on the crosslinking rate, as crosslinks stabilize the channels, and this is a possible mechanism since earlier chapters concluded that HCl definitely affects the crosslinking process. Analyzing the effect of channel presence on the diffusion process, let us use another structure to represent the channel.

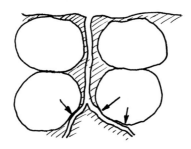

Fig.5.7. Channel effect.

Suppose we have a channel formed like that in Fig.5.7. The channel ducts the HCl from inside the sample directly to its surface; therefore, the process of HCl release from this particular free space would be easier (not even necessarily connected with real diffusion) as this chamber would have lower a HCl concentration, which should certainly cause diffusion to change direction (otherwise it would be directed to the surface) as the concentration gradient is the motive force of the diffusion process. This discussion should be especially adequate to describe the process of HCl emission in advanced stages. Furthermore, due to the rather slow diffusion through envelopes, it can be predicted that HCl will accumulate inside the envelopes and pressure build up due to which grains should increase their dimensions. Such a process can be related to the structure´s swelling; this, again, will be reflected in changes of free spaces due to the larger dimensions of the grains.

Let us relate these processes to the reaction with stabilizer as we started to in the beginning. We should notice that a slow, undirected diffusion of HCl is preferable for efficient reaction with stabilizer, since there is the highest probability of such a reaction occurring, and stabilizer is used most uniformly. Any channel formation restricts the movement of one reacting component, while the other is not sufficiently mobile to migrate efficiently to the center of a possible reaction. If we take the alternative of another stabilizer able to penetrate envelopes, then on one hand we should have a more effective stabilization process as the free way of gas is shorter and there is a higher probability of direct reaction between stabilizer and excited elements of the polymer chain. On the other hand, there is the danger that if reaction products of the stabilizer are able to catalyze the degradation process, then they will serve finally as degradation centers (e.g., zinc sensitivity). It was remarked earlier that the most important attribute for initial color stability is effective HCl binding. From this explanation it is evident that only stabilizers which can penetrate inside the grains are able to act as good initial color stabilizers.

Summarizing the above discussion, the reader should be warned that for lack of adequate data and studies in this field, many explanations given here are sometimes based more on speculations than confirmed facts. The author, after reviewing the literature, had two options at hand : either to satisfy scientific purity and scrap the chapter for which studies were not available, or to take the risk of being accused of drawing premature conclusions based on related fields of study or pure intuition. As is evident from the above pages, the more risky way was taken, and the reason was a strong belief that we are not going to succeed in further understanding if serious research is not done on degradation and stabilization kinetics in relation to structural effects. Possibly many conclusions will prove inadequate in the future. Therefore, it will only be necessary to re-write some pages when new directions of studies are found.

REFERENCES

1. J. Wypych, **Proc. ACS** Div. **Polym. Mat.,** 52(1985)545.

2. J.H. Hildebrand in **Viscosity and Diffusivity,** John Wiley & Sons, New York, 1977.

3. P.L. Shah in **Encyclopedia of PVC,** Chapter 22, Marcel Dekker, New York, 1977.

4. J.J. Pena, A. Santamaria and G.M. Guzman, **Eur. Polym. J.,** 20(1984)49.

5. **German Pat.,** 26 52 328.

6. J. Lyngaae-Jørgensen, **Pure Appl.** Chem., 53(1981)533.

7. C.L. Sieglaff, **Pure Appl.** Chem., 53(1981)509.

8. F.N. Cogswell, **Pure Appl.** Chem., 52(1980)2031.

9. E.M. Katchy, **J.** Appl. **Polym.** Sci., 28(1983)1847.

10. G. Ceccorulli, M. Pizzoli, M. Scandola, G. Pezzin and G. Crose, **J. Macromol. Sci.-Phys.,** B20(1981)519.

11. D.M. Gezovich and P.H. Geil, **Int. J. Polym. Mat.,** 1(1971)3.

12. P.K.C. Tsou and P.H. Geil, **Int. J. Polym. Mat.,** 1(1972)233.

13. V.P. Lebedev, B.P. Shtarkman and T.L. Yatsinina, **Tr. Khim., Khim. Tekhnol.,** 3(1972)135.

14. C. Singleton, J. Isner, D.M. Gezovich, P.K.C. Tsou, P.H. Geil and E.A. Collins, **Polym. Eng. Sci.,** 14(1974)371.

15. D.L. Tabb and J.L. Koening, **Macromolecules,** 8(1975)929.

16. J.C. Scaiano, E.A. Lissi and L.C. Steward, **J. Am. Chem. Soc.,** 106(1984)1539.

17. A.R. Amer and J.S. Shapiro, **J. Macromol. Sci.-Chem.,** A14(1980)185.

18. G. Lèvai and G. Ocskay, **J. Macromol. Sci.-Chem.,** A12(1978)467.

19. E.D. Owen, I. Pasha and F.Moayyedi, **J. Appl. Polym. Sci.,** 25(1980)2331.

20. M.C. Ball and M.J. Casson, **J. appl. Chem. Biotechnol.,** 28(1978)765.

21. J. Verdu, **J. Macromol. Sci.-Chem.,** A12(1978)551.

22. **ISO-Technical Report** 4616-1980.

23. M. Doi and S.F. Edwards, **J. Chem. Soc., Faraday Trans. 2,** 74(1978)1789.

24. M. Doi and S.F. Edwards, **J. Chem. Soc., Faraday Trans. 2,** 74(1978)1802.

25. J. Klein, **Macromolecules,** 14(1981)460.

26. P.T. Gilmore, R. Falabella and R.L. Laurence, **Macromolecules**, 13(1980) 880.

27. B.A. Smith, **Macromolecules**, 15(1982)469.

28. A.Z. Akcasu, **Macromolecules**, 15(1982)1321.

29. J.G. Curro, R.R. Lagasse and R. Simha, **Macromolecules**, 15(1982)1621.

30. M. Schmidt, W.H. Stockmayer and M.L. Mansfield, **Macromolecules**, 15(1982).

31. H.T. Lin, J.L. Duda and J.S. Vrentas, **Macromolecules**, 13(1980)1587.

32. J. Wypych, **ACS Polym. Preprints,**) 26(1985)122.

CHAPTER 6
ANALYTICAL METHODS USED IN PVC DEGRADATION STUDIES

6.1. Sample choice and preparation

The problem of sample choice and preparation has already been indicated in several parts of the present monograph, whereas general practice and observations are critically reviewed here. Sample selection consists of two important factors:

1. The method of polymer synthesis.
2. The method of specimen preparation.

Our studies conducted in this field are designed mainly to facilitate practice; therefore, the industrial products of vinyl chloride polymerization seem to offer the best choice. At the same time, all industrial polymers contain additives, and they are a mixture of a broad spectrum of molecular weights, which arouses concern that they may interfere in studies being conducted. We, therefore, have a choice of three other methods:

- polymer purification,
- polymer fractionation,
- polymerization under laboratory conditions.

Polymer purification is done either by Soxlet extraction with methanol and hot water washing or by tetrahydrofuran dissolution and methanol precipitation and washing. The first method seems to affect the initial properties of polymer less, since only limited extraction of low molecular fractions of polymer were reported, while the second method leaves behind trace quantities of solvent (THF), which may participate in a further photooxidation process as discussed in Chapter Three. Polymer fractionation has a similar impact on results since fractions are usually obtained by gradual precipitation of polymer from its solution by periodic dosing of non-solvent. As a result of this process, one can obtain samples of a narrow range of molecular weight distribution.

Laboratory synthesis may lead to various effects, depending on the kind of studies being conducted. Most frequently, this method is used for polymer obtained by UV irradiation, in order to limit the additives in the sample used for further studies. One should realize that such synthesis leads to polymer that is not directly comparable with industrial products, and hence the conclusions can be atypical.

In most recent studies within the last five years, 90% of polymer samples used for further investigations were industrial products not subjected to any treatment before they were formed to the intended specimen shape. Only in special cases was the polymer applied obtained in laboratory synthesis. This includes studies on tacticity effect and studies of labile structures wherein a larger variety of tacticity is needed than that offered in industrial products. Purified samples of polymer were used occasionally in studies of thermodegradation kinetics, stabilization and photolysis. The above discussion shows that the authors confined themselves to a more practical approach. It is interesting that in many research undertakings the same industrial polymer was used. Breon 121 is the most frequently used standard polymer type, but some national standard types of polymer are used for studies as well.

Considering the method of specimen formation we have a wider variety of techniques:
- film cast from solution,
- polymer solution,
- powdered polymer,
- blend of polymer with other ingredients,
- polymer coated with ingredients in solution,
- ingredients mixed with polymer in mortar,
- extruded specimen,
- foil manufactured from plastisol,
- compression molded sheet,
- calendered sheet,
- IR die pellets.

One cannot say that any particular technique is especially convenient and appropriate, since the form of the specimen depends in the first place on the analytic method used in the course of studies. This is why about 25% of specimens used in studies were obtained in the form of solution or the film cast from solution, usually from tetrahydrofuran. The method used in investigation demanded high transparency, which cannot otherwise be achieved without advanced changes in the material studied that are caused by thermal degradation. At the same time, every investigator using samples treated with solvent was probably aware that the traces of solvent left in the specimen affect the course of the photooxidation reaction; therefore, in these studies the type of analytic method played a predominant role. The use of solution, polymer cast from solution or polymer precipitated from solution is the most risky method of sample preparation, and therefore, it should be used only when other methods fail to provide the functions demanded. In this case there are several solvents in common use, including: tetrahydrofuran, dichloroethane, 1,2,4-trichlorobenzene, dimethylformamide, cyclohexanone hexamethylphosphortriamide and pirydine. From the point of view of its effect on initial polymer properties, cyclohexanone should be the first choice; unfortunately, it has too high a boiling temperature and viscosity, which rule out its use in the majority of cases. Tetrahydrofuran is the most commonly used solvent, even though it is the most inadequate due to its tendency to oxidation and peroxide formation. The most important factors to consider are related to solvent purification (1), but at the same time one should bear in mind that further use of a specimen in the presence of oxygen has a detrimental effect on its condition.

The original powder form of PVC is the most appropriate for further studies, but in the thermal degradation test, the size of the grain plays an essential role. In consequence, the sample should be repeatedly selected for a narrow range of grain size or sieved prior to use. In some studies samples were prepared using an IR die for pressing pellets. Here the thermal effect is minimal; therefore, the sample could be regarded as similar in properties to the original form. The disadvantage of this method is that the sample is fragile and lacks transparency. Another relatively low heat-treated sample could be formed from plastisol in temperatures much below the gelation point (about 373K), but this technique is limited only to studies including plasticizers.

Film calendering, extrusion and compression molding of polymer sheets are also frequently applied, as about 20% of samples have been prepared this way for various recent studies. It is understood from the principles of these methods that thermal changes are introduced into the material in its initial form. In order to achieve replicable results, a strict regime of processing is applied in which temperature is controlled, as is shear rate, which indirectly is also related to the level of heat treatment the sample obtains. Although these methods are the most process oriented, they may introduce quite advanced changes since it is not exceptional for the sample to be processed for 10 min at 453K.

A group of techniques are applied when polymer is studied in the presence of additives, as for example, in stabilization research. In this case, even distribution of additives is very important. Usually the additive has a different bulk density; so mixing in a mortar alone may prove inadequate. Blending the mixture in high speed mixers with temperature increase (up to 393K) gives good results in the distribution of additive, but some thermally induced changes may occur which affect further results. The most even coating of polymer can be achieved by dissolving the additive in solvent that does not interact with polymer. But such an approach should not include the polymer solubility alone. In one study the stabilizer was dissolved in ketone, which is known to be able to interact with polymer, and such practice is too risky so far as research output and conclusions are concerned.

From the above discussion, one can easily come to the conclusion that sample preparation alone can be at least as important for the results obtained as the measuring technique. Therefore, the method of sample preparation should not only be cautiously selected but the results scrutinized, investigating the possible effect of the method on studies being conducted.

6.2. Equipment for thermal degradation measurements

Measurements of PVC thermal degradation are most important for the evaluation of polymer thermostability; therefore, the development of new methods is still of great interest and importance. One might expect a detailed literature on this subject. In practice, a great variety of methods are known, but only a few papers give details of equipment construction, so practically every laboratory uses its own model. This is why we found so many contradictions in discussing the data in previous chapters, especially when the measurement of PVC thermal degradation, which is very simple in principle, is complicated by numerous parameters.

The simplest methods, designed some years ago, such as determination of initial period by Congo red paper and solution (2,3) or by $AgNO_3$ solution (3), are not used any longer. The methods of sample color change under heat treatment have only industrial applications (4,5). The same is true of the method for measuring the electrical conductivity of a sample (6-8). Pressure measurement of gaseous products emitted by the sample is not used because of the interference of HCl´s catalytic effect (9,10).

The most commonly used methods are those in which the sample is subjected to isothermal heating, and evolved hydrogen chloride is monitored after absorption in distilled water. HCl determination is done by one of these methods:

- potentiometric titration (11,12),
- argentopotentiometric titration (13-16),
- pehametric (17-21),
- conductometric measurement (5,11,22-29).

The last method, conductometric measurement, is by far more popular than any other at present.

There is continuous interest in using thermogravimetric analysis (TGA) and differential thermogravimetry (DTGA), although practically they could be applied to samples free of volatile substances, e.g., plasticizers (30-31). The original method of HCl monitoring was used in Australia (36), where the IR spectrophotometer was applied to detect HCl evolvement at 2963 cm^{-1}. In another new development, Guyot and Bert (28) applied differential conductometry.

The detection limit for potentiometric and argentopotentiometric titration was quoted as 10^{-2}% of total HCl in the sample. In the case of thermogravimetric measurement, the same value equals 1.2×10^{-1}% for a 5 mg sample. For the pH-metric method, the given detection limit of 10^{-4}% seems to be overestimated (18). In the case of the conductometric method,

Abbas (29) also claimed to obtain a detection limit of 10^{-4}%. If we use a conductivity meter that can measure 0.01 S in a sample of 0.1 g, we can detect 3.3×10^{-3}% of HCl, which is equivalent to 9.3×10^{16} HCl molecules evolved, meaning that roughly every tenth chain of average molecular weight of 8×10^4 had split off one molecule of HCl before we were able to detect it.

If one takes into consideration the mechanisms we want to observe, one has to look for the best possible execution of the rather rough method we are using at present. The amount of sample taken for measurement is quite important. By increasing the sample amount by the order of ten, one may increase sensitivity by the same value, but at the same time thermal balance, i.e., isotermic conditions, will take longer to achieve. Even more important is sample thickness. As the surface area on which we can place the sample is usually limited, by using a bigger sample we also use a thicker layer, and because polymer, under conditions of measurement, is of high and increasing viscosity, HCl escapement gradually becomes more difficult. Unless taken by carrier gas into the conductivity cell, HCl stays in the sample, which leads to uncontrolled conditions of degradation since there is no method yet proposed to separate the effect due to heat treatment from continuously changing extent of the catalytic influence of hydrogen chloride on the degradation rate.

The other essential factor concerns the material from which the reaction vessel is made. Contact between the sample and the reaction vessel´s surface may produce an additional unexpected effect, i.e., concerted elimination or the process of initiation and termination of HCl split off, which is catalyzed by reactor walls. This might cause a lot of experimental errors, as in most cases experimental vessels are composed of Pyrex glass, which may contribute to the wall-catalyzed effect. Pyrex glass reactors might be prepared for degradation studies of halides by formation of carbonaceous coatings. For degradation at temperatures below 583K, PTFE-coated reactors provide the most inert surface (37). One should remember that Teflon has a very low thermal conductivity coefficient, and so the thickness layer should be minimal. Teflon should also be used to help clean the surface before another sample is introduced.

The dimensions of the degradation vessel are also important. The size of the reaction vessel does not practically affect the character of carrier gas flow, as the value of Reynolds´s number is very low ($\sim 1 \times 10^{-3}$); therefore, carrier gas follows the pattern of laminar flow. One should expect that under such conditions of flow, the heat exchange rates would be very poor. The volume of the reaction vessel is important because, together with carrier gas flow rate, it determines the rate at which the atmosphere surrounding the sample is exchanged, which is important for diminishing the catalytic effect of HCl mentioned above. Under practical conditions used at present, one might suggest the following range of carrier gas replacement frequency: 5-20 total exchanges of carrier gas per minute. One should keep this value as high as possible, not only to diminish the catalytic effect, but also to achieve better mixing of gas inside the reactor and faster transportation of HCl to the conductivity cell, thus providing for its faster detection.

Isothermic conditions under which the sample is degraded are probably the most difficult to achieve, and they contribute to most data error. Vymazal (38) has described a method of calculation for the correction factor for non-isothermal conditions. He has also given the values that characterize the rate of temperature increase in the sample, as shown in Table 6.1.

Assuming that a tolerance of 1K can be achieved at present, we could arrive at this range only after 8 min, which means that, depending on PVC type, if degradation is measured at 453K, 4×10^{-3}-2.6×10^{-2}% of the sample would already have been degraded. In other words, 2.2×10^{17}-1.3×10^{18} of HCl molecules were produced, which means that every chain had already formed one double bond. Wang (21) claims that the tolerance of 1K is reached 4.5 min after the sample is introduced, while Abbås (29) claims that 2 min is needed to approach a temperature of 463K by 1K difference. Based on these data, it seems important to discuss the problem of isothermal conditions in detail. The conditions of sample heating once the sample has been introduced to the reaction vessel can be illustrated by the following diagram (Fig. 6.1).

Table 6.1.

**Sample temperature increase during degradation measurement for
0.2 g of PVC powder. Carrier gas flow rate: 330 cc/min.**

(Data from Ref. 38.)

t, min	T, K	t, min	T, K
0	293.0	7	451.5
1	373.0	8	452.4
2	405.0	9	452.6
3	428.0	10	453.0
4	443.6	12	453.1
5	448.0	14	453.0
6	450.4		

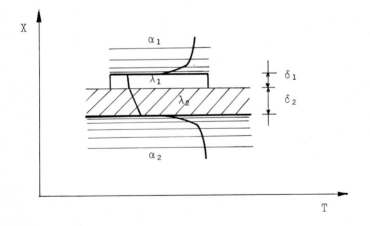

Fig.6.1. Sample heating diagram.

The sample is heated from the top by gas, and the process of
heat exchange is limited by values of convection heat transfer
coefficient α_1 and by the conductivity of sample λ_1. From the
bottom, the sample is usually heated by hot oil, and the process
is controlled by convection heat transfer coefficient α_2 and
conductivities of the glass wall and the sample λ_2 and λ_1,
respectively.

In order to obtain conditions closest to isothermal conditions of sample degradation, we should try to achieve the highest values for α_1, α_2, λ_1, and λ_2, δ_1 and δ_2 should be as low as possible, and $T_1 = T_2$. We are not able to limit the values of λ_1 and λ_2 as PVC conductivity depends on sample that is measured and α_2 is the conductivity of glass. A change in α_1 can only be effected by a very drastic change in the carrier gas flow rate, which is unrealistic under present conditions. Also δ_1 cannot be controlled adequately, as under measuring conditions the sample becomes a plastic material and the layer thickness will depend largely on its viscoelasticity and the surface tension of PVC melt. What we can do is to limit the α_2, δ_2 and $(T_1 - T_2)$ difference. The coefficient is given by the function below:

$$\alpha = f(w, d, L, \eta, c, \lambda)$$

where:

w – rate of flow of fluid,

d, L – linear dimensions of the reaction vessel,

η – fluid viscosity,

c – specific heat of fluid,

λ – conductivity of fluid.

Therefore, we should increase w and λ, and decrease c and η. Unfortunately, the practical values of w in an oil bath cannot be increased drastically enough to achieve the change of flow character, i.e., from a laminar to a turbulent one. One should, therefore, try to experiment with the oil used in the oil bath in order to optimize the values of λ, c and η. Another possibility was used in the author's laboratory (39). As convection in comparison to conduction is a rather slow process of heating, one may foresee that if the bottom of the sample is to be heated through conduction, the results should be much better. Fig. 6.2. shows the details of design of such equipment. The element shown in Fig. 6.2 is part of the measuring set shown in Fig. 6.3. The glass reaction vessel has a diameter of 0.2 mm less than the internal diameter of the brass heating block; therefore, the glass vessel and brass block are in almost direct contact.

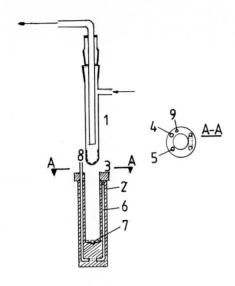

Fig.6.2. The construction of the degradation vessel. 1 - glass degradation vessel, 2 - brass block, 3 - PTFE insulating ring, 4 - carrier gas inlet, 5 - carrier gas outlet, 6 - gas heating channel, 7 - silicon oil, 8 - PTFE coating, 9 - thermocouple.

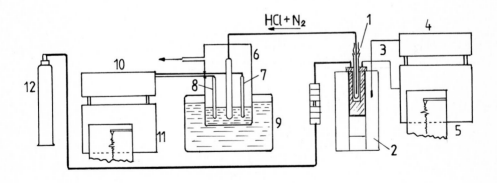

Fig.6.3. PVC degradation measuring apparatus. 1 - degradation vessel, 2 - furnace, 3 - thermocouples, 4 - temperature regulator, 5 - temperature recorder, 6 - conductivity vessel, 7 - conductivity probe, 8 - temperature sensor, 9 - thermostat, 10 - conductivity meter, 11 - conductivity recorder, 12 - carrier gas cylinder, 13 - rotameter.

To improve the conditions of heat exchange still further, a small amount of silicon oil is introduced inside the brass block to seal both surfaces. A very thin oil layer transfers the heat primarily by conduction as "wall layer" is diminished. Conditions of heat exchange can be further improved if oil is replaced by mercury, but use of the latter is limited by safety precautions.

It is interesting to compare the equipment presented above with the other equipment in use. Figs. 6.4 and 6.5 show well designed equipment used in Sweden (29).

Both measuring units differ in several respects, but the most important differences concern the method of sample heating. In Abbås´ design, gas is heated in a separate glass coil that has a surface area about eight times larger than in the author´s design in which the channel is mounted directly inside the brass block.

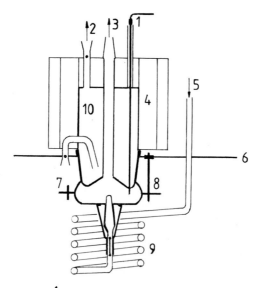

Fig.6.4. Degradation vessel. 1 - thermocouple, 2 - polyene glycol outlet, 3 - carrier gas outlet, 4 - asbestos insulation, 5 - carrier gas inlet, 6 - brass plate covering thermostat, 7 - polyene glycol inlet, 8 - reaction vessel holder, 9 - glass coil for preheating the carrier gas, 10 - heating jacket. (Modified from Ref. 29.)

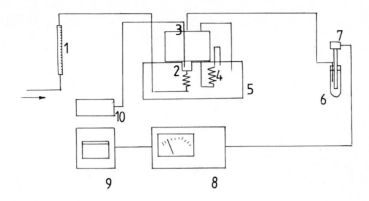

Fig.6.5. Apparatus assembly. 1 - rotameter, 2 - lower part of reaction vessel, 3 - thermocouple, 4 - heat exchanger connected to upper part of reaction vessel, 5 - thermostat, 6 - conductivity measuring cell, 7 - probe, 8 - conductivity meter, 9 - recorder, 10 - digital voltmeter. (Modified from Ref. 29.)

A few experiments were done to compare both methods of gas heating. In one case, gas was heated in brass block; in the other, in a 2 m long glass coil of 8 mm diameter. The temperature was measured close to the bottom of the glass reaction vessel by sensor, which was introduced in the first case into the brass block, and in second, into the oil bath. In both cases the temperatures of the block and the oil were controlled. Results of measurement are shown in Fig. 6.6. Clearly, much better uniformity of gas temperature, compared with that of the heating medium, can be obtained in a metal heater. Another set of experiments was performed to investigate the time needed for equilibrium conditions to be achieved after a sample of 0.1 g was introduced. Comparison was made in two different ways. In one case, a sample was introduced to a reaction vessel kept in a heating block or oil bath, and in the other a reaction vessel was taken away from the heating medium for 1 min and introduced to heating media again together with the sample. Figs. 6.7 and 6.8 show the results. As might be expected, equilibrium is achieved earlier in the metal block heating element.

240

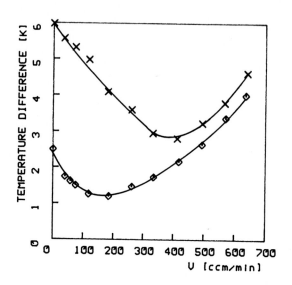

Fig.6.6. Temperature difference of heating media and gas versus gas flow rate (V). × - block heating element, ◇ - oil bath (39).

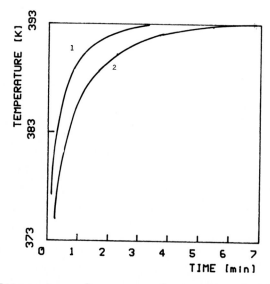

Fig.6.7. Temperature increase after introducing the sample and the reaction vessel. 1 - block heating, 2 - oil bath heating (39).

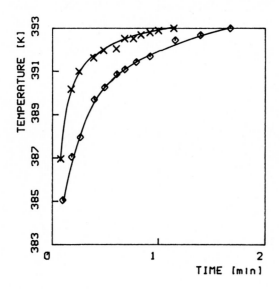

Fig.6.8. Temperature increase after introducing the sample of 0.1 g (+ - block heating; ◇ - oil bath heating) (39).

In order to complete our discussion of the problems connected with measuring the thermal degradation process of the PVC sample, one should mention briefly some other important and obvious points:

1. The distilled water used in conductivity measurements should be of the highest possible quality.
2. The residual conductivity of water after saturation with inert gas should be constant before measurement starts and at least less than $0.5\,\mu S$.
3. Thermostating of water and the use of automatic thermal compensation should serve to obtain reliable data.
4. A standard curve for at least 30 points of different dilutions with statistical approximation should be carefully prepared.
5. Sample loading time should be diminished.
6. One should not allow a higher error than 0.2% in sample weighing.
7. All glassware and tubing should be made out of inert material and cleaned and dried after each measurement.

8. Carrier gas should be transferred through the system by overpressure rather than by using vacuum pumps, because in the latter case, during the time of sample loading, water is being saturated by carbon dioxide, which influences conductivity measurement.

9. Removing part of the sample during the degradation process must inevitably produce errors.

There is in use, although not very commonly, a method of measuring thermal degradation by an objective test of color change using CIE tristimulus values (40,41). The objective color measurement is broadly applied in many fields, e.g., the textile industry, where color poses the essential quality factor. This method was reapplied to PVC samples being degraded under constant temperature conditions but at varying times. Sample color change is measured by a yellowness index (YI) given by the equation:

$$YI = 100\frac{1.28X - 1.06Z}{Y}$$

where:

X,Y,Z - CIE tristimulus values.

Samples which have YI=0 did not change color, those having YI>0 are discolored towards yellow, and those having YI<0 are discolored towards blue.

The best application of this method is in measuring initial color change, since the higher degradation stages may not be monitored with initial sensitivity. Using this method, one should always have a standard sample thickness and a standard background on which the sample is placed for measurement, bearing in mind that practically every sample is transparent; therefore, both parameters affect the measured value. This method has one serious shortcoming, since it can be operated only on a periodic basis. The results would be more useful if they were comparable with results of other measurements, e.g., HCl emission rate, but this is not possible at the moment since the method does not allow for continuous measurement. On the other hand, it is already a step in the right direction, since we are able to observe changes by more objective instruments than the human eye.

6.3. Molecular weight determination

Four methods are applied in PVC molecular weight determination. The simplest method includes determination of the viscosity of the PVC solution, the result of which is then expressed in the form of molecular weight calculated from the Mark-Houwink equation. PVC is usually dissolved in cyclohexane or tetrahydrofuran and measured at 298 or 303K (42-45). Some notice should also be given to the membrane osmometric method, which was applied by some investigators (43,46) for cyclohexane solutions measured at 298K.

Gel permeation chromatography is by far the most popular method for measuring molecular weight and molecular weight distribution of PVC samples. Waters Associates GPC models 200, 201 and 244 are usually used for this purpose (42,44,47-50). The sample, dissolved in tetrahydrofuran in a concentration from 0.2 to 0.4%, is measured at 298K on a set of Styragel columns, usually with the following exclusion limits: 5×10^6, 7×10^5, 1.5×10^5, 1.5×10^4 and 1.5×10^3 $\overset{o}{A}$ (49). For samples of lower molecular weight a 500 $\overset{o}{A}$ column is used. The usual flow rate of solvent (peroxide-free THF) is in the range of 1-2 ml/min. The columns are calibrated with the polystyrene standard, and the results are frequently calculated by Drott´s computer program (51). The method gives not only the value of average molecular weight but also its distribution.

In the case of interest in low molecular weight oligomers of vinyl chloride, steric exclusion chromatography is used (52,53). This method can help to separate molecules of molecular weight below 1000. PVC in solution in THF at a concentration of 0.2-0.3% is subjected to chromatographic analysis on Sephadex LH 60.

From the above review it is evident that the methods of PVC molecular weight determination are well studied, based on objective measurement and broadly applied. It seems that this area of study has a solid foundation. The only concern about replicability of results might be associated with the method of THF purification, since the presence of peroxides should certainly affect the result. In some cases the results might also be misleading, either if PVC is not dissolved properly, which would lead to the association of molecules, or if the

244

sample contains polymer that is partially insoluble due to former treatment.

6.4. Ozonolysis

PVC ozonolysis is applied to study the internal unsaturations in the polymer chain. Since such unsaturations are usually not numerous, it is difficult to study them. Eventual oxidation of unsaturation present at chain ends would practically not affect the molecular weight of the polymer, which is the opposite of the case when unsaturation is in the chain´s internal sections. These observations contributed to the development of ozonolysis, which is widely used at the moment for studies of internal unsaturations in polymer and polyene distribution (49,50,54).

Michel (55) reported studies on various PVC types at varying conditions in order to establish the course of ozonolysis. Figs. 6.9 and 6.10 show the effect of ozonolysis temperature on process efficiency.

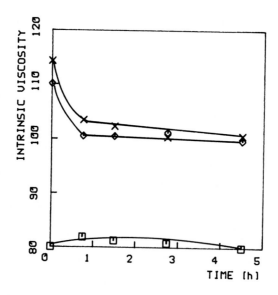

Fig.6.9. PVC ozonolysis at 253K. × - suspension PVC, ◇ - emulsion PVC, □ - bulk PVC. (Modified from Ref. 55.)

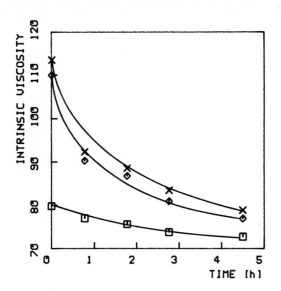

Fig.6.10. PVC ozonolysis at 293K. × - suspension PVC, ◇ -
emulsion PVC, □ - bulk PVC. (Modified from Ref. 55.)

Bulk PVC is evidently degraded only at higher temperatures, and
the temperature has an appreciable effect on the rate of
ozonolysis. In light of this work, it is surprising that
ozonolysis sometimes is carried on at 253K (54) since the above
results show that some samples may remain intact. In other
studies ozonolysis is performed at room temperature (50) or at
273K (49) in chloroform media containing some methanol. There is
some disagreement among authors (50,55) as to the effect of
ozonolysis temperature on the occurrence of side reactions.

Summarizing these studies, one should say that, although the
idea behind ozonolysis is very appealing and the method itself
presents a very valuable tool in PVC degradation studies, at the
same time the conditions of ozonolysis should be studied in more
detail, as otherwise the results may not be relevant to the
subject of investigation.

6.5. Chemical modification in studies of PVC structure

In the search for methods of studying irregularities in the PVC
chain, chemical modification has been found to be an important

analytic technique. We have already discussed such techniques, since ozonolysis may also be regarded as a means of chemical modification. Other reactions such as reduction, chlorination, bromination, acylation, phenolysis and copolymerization have also been found useful in these studies.

Caraculacu (56,57) found that the labile structures of the PVC chain can be replaced by phenol, forming the following structures:

~CH=CH-CH~$_2$

$$\text{OH (para-substituted phenyl ring)}$$

~CH=CH-CH-CH~$_2$

$-CH_2-CH-CH_2-\overset{\displaystyle}{C}-CH_2-\overset{}{CH}-$
with OH, CH$_2$, CHCl, Cl substituents

which means that both the chlorine in the vicinity of the double bond (internal and terminal) and tertiary chlorine are replaced by phenol rests while chlorine in regular units is not. This gives an opportunity to distinguish between labile and normal chlorine by spectral methods.

In order to perform this reaction, the PVC sample is heated at 333K for 96 hrs in an excessive amount of phenol; then phenol is washed out by methanol, and the PVC sample is dissolved in tetrahydrofuran and precipitated by the addition of methanol at 278K. Finally, the sample is dried in a vacuum and subjected to spectral analysis (58).

This method, when combined with other modes of chemical modification, e.g., bromination, can give additional data (59,60). The bromination reaction affects double bonds:

$$-CH=CH-CH- + Br_2 \longrightarrow -CH-CH-CH-$$
$$\quad\quad\;\; | \quad\quad\quad\quad\quad\quad\quad\;\; |\;\;\; |\;\;\; |$$
$$\quad\quad\;\; Cl \quad\quad\quad\quad\quad\quad\quad Br\;\; Br\;\; Cl$$

It is evident that chlorine ceases to be in the vicinity of the double bond; therefore, it may not be a labile structure any longer, as Buruiană (57) confirmed. If phenolysis is done after bromination, it will replace only the tertiary carbon-bound chlorine. From the difference we can estimate the concentration of double bonds in the PVC chain (both internal and terminal).

For the bromination reaction, PVC is dissolved in dichloroethane at 333K, then mercury acetate solution in acetic acid is added as a reaction catalyzer and, finally, bromine solution in dichloroethane is introduced into the reacting mixture. Reaction is performed at room temperature, in darkness, for 24 hrs. The bromination technique can be used for a direct reading of double bond concentration, since an excessive amount of bromine can be reacted with potassium iodate and this titrated potentiometrically with thiosulphate solution (60). Scott (61) applied a similar method to determine the total unsaturations in the course of thermal degradation. In his studies samples were reacted with iodine in excess and the remainder titrated with $Na_2S_2O_3$. Guyot proposed an alternative method of allylic chlorine determination with thiophenol:

$$\sim\!CH\!=\!CH\!-\!\underset{\underset{Cl}{|}}{CH}\!\sim \;\;+\;\; \underset{}{\bighexagon}\!-\!SH \;\longrightarrow\; \sim\!CH\!=\!CH\!-\!\underset{\underset{|}{}}{CH}\!-\!S\!-\!\bighexagon \;\;+\;\; HCl$$

This reaction has the advantage that it allows one to monitor simultaneously thiophenol consumption, HCl emission and ring grafting onto the PVC. According to Guyot (62), this reaction takes place only with allylic chlorines, while those bound to tertiary carbon atoms remain intact. The reaction is carried out in dichloroethane solution at 333K for 30 hrs, under nitrogen which carries HCl to the absorption vessel. Thiophenol concentration can be monitored by coulometric titration, but there are two problems with this method. First of all, it is not clear which double bonds play a role in the reaction. Secondly, the method is confirmed for the reaction of PVC low molecular weight models. Although this method was applied later to PVC, there is still no direct evidence of how it compares with other methods.

The other experiments with sulphur-containing compounds lead us to different conclusions than those given by Guyot (62). Starnes (63) shows the effect of pretreatment of PVC samples by dibutyltin bisthiododecane (Fig.6.11).

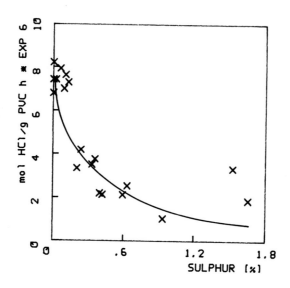

Fig.6.11. Dehydrochlorination rate versus sulphur content in chemically modified PVC. (Modified from Ref. 63.)

Fig. 6.11 clearly shows that with the increase in sulphur concentration, the dehydrochlorination process slows down, which is expected due to the reaction with labile chlorine. Other studies by the same author reveal that, first of all, the reaction is to be written in the following form:

$$-CH_2-CHCl- + 2RSH \longrightarrow -CH_2-CH_2- + RSSR + HCl$$

Secondly, there is no discrimination so far as the thiol type is concerned (aromatic or aliphatic), and the reaction yields 29% of the reduced product. As a matter of fact, Starnes has done his experiments at a temperature higher by 100K than Guyot (62), who felt that temperature above 353K might lead to further dehydrochlorination. At the same time, it seems doubtful that the reaction can be so temperature-dependent and yet be applicable for analytic purposes.

Carrega (64) has developed a procedure for PVC reduction. The idea of his method is to replace all chlorines in the PVC chain in order to be able to use the technique broadly applied for polyethylene, based on NMR analysis, for determination.

According to Carrega's method, the PVC sample is reduced in an autoclave under nitrogen of high purity (<30 ppm O_2) with $LiAlH_4$ in tetrahydrofuran. The reaction is allowed for seven days at 373K. Then, PVC is first separated from reagents by pouring the mixture into ice-cooled 6 N hydrochloric acid, then dissolved in decalin, precipitated in ice-cooled acetone and dried at 323K for two to three days. This method is known to be able to introduce double bonds into the polymer chain, destroy branch points containing tertiary chlorine, and yet it does not yield a completely-reduced product (62). This is why the new reagent, Bu_3SnH or Bu_3SnD, was employed by Starnes for this type of study (63). The reaction proceeds according to a free-radical chain process:

$$RCl + Bu_3Sn^{\cdot} \longrightarrow R^{\cdot} + Bu_3SnCl$$
$$R^{\cdot} + Bu_3SnH(D) \longrightarrow RH(D) + Bu_3Sn^{\cdot}$$

with no evidence of competitive reactions, and it leads to a completely reduced product that serves as material for further study of branching points by NMR. Based on this method, PVC in 2-methyltetrahydrofuran solution is refluxed at 353K with tri(n-butyl)tin hydride and azobisisobutyronitrile for 24 hrs under nitrogen atmosphere. A sample of solid polymer is separated by filtering and washed by 2-methyltetrahydrofuran and methanol, then Soxlet extracted with methanol. The final polymer usually contains 2-4% chlorine. This sample is again reduced with tri(n-butyl)tin hydride, this time in xylene solution for another 24 hrs at 363K under nitrogen. Afterwards, the polymer, when precipitated with methanol and washed and dried, contains 0.0-0.4% chlorine (65).

Another attempt to collect data from chemical modification studies is presented by Braun (54), who prepared copolymers of vinyl chloride with phenylacetylene and dimethylester of 2-butyne dioic acid in order to study the effect of unsaturations. In other research, Braun (66) presented CO copolymerization in order to discuss the effect of carbonyl groups on the dehydrochlorination rate of the PVC chain.

Summarizing chemical modification studies, one may observe that some are a kind of alternative to dehydrochlorination measurements, which experience analytic difficulties as discussed earlier. Although new, these methods are well studied and helpful for overcoming the problems of dehydrochlorination measurements. Other studies concentrated on labile structures in the PVC molecule have recently contributed to a far better understanding and may be a good sign that in the near future we shall be able to obtain sufficient data on this important subject.

6.6. Spectroscopic methods in PVC studies

6.6.1. UV-visible spectrophotometry

UV-visible spectrophotometry is one of the most prominent methods in PVC thermal degradation studies since polyenes formed in the process absorb in this range. Fig. 3.27 and Table 3.4 show typical data obtained from UV-visible analysis. Since the method is common knowledge, we limit our discussion to sample preparation.

There are basically two forms of sample used for UV-visible spectra determination, i.e., solution, and film cast from solution. PVC solution used for these studies is usually prepared in tetrahydrofuran (36,42,49,50,67) in concentrations from 0.75 to 4 g per liter. Unlike other authors, Millán (43-47) used 2-4 g/l PVC solutions in hexamethylphosphortriamide, and Scott (61) used a 7% PVC solution in methylene chloride. The solution method has two important limitations. First, in the course of PVC degradation, samples become less soluble and part of the polymer is lost to determination; secondly, the solvent plays an essential role in the degradation process if handled in the presence of oxygen. This is why all authors have stressed purification methods.

Studies of UV-visible spectrum with a sample in the form of film are still rare (68-70). In every case, the film was prepared by casting from THF solution; therefore, the above remarks apply here as well, but the effect of solvent can be diminished if the proper procedure is followed. Film thickness varies from study to study. Vymazal (68) used films of 55-145 μm, recalculating results obtained to the thickness of 65 μm. Davidson (69) used films of 30±5 μm, while Verdu (70) analyzed the problem of film thickness in detail, and his conclusions are given below. Vymazal (68) and Davidson (69) did not suggest that the films used for their studies were processed in any special way, except being cast from solution.

Verdu´s method (70) seems to be the most carefully chosen and studied in detail, and it might be proposed as an example of research technique for UV-visible studies. Film was prepared from a solution of 5 g PVC in 100 cm^3 of THF. The evaporation rate of the solvent was controlled so as to obtain the expected optical clarity of film. The film had to have at least 80% transmission in the near UV. THF was stabilized with an antioxidant (2,6-tert-butyl-4-methylphenol) that prevents peroxidation but at the same time can remain in the sample. After preparation, the film was extracted in Soxlet apparatus by ether in order to remove traces of THF and the antioxidant. The process was completed by diethyl evaporation at 313-333K under vacuum and control of the sample for THF and presence of antioxidant. Residual content of THF can be checked by IR since it absorbs at 1060 cm^{-1}. If antioxidant is present in the sample, it will absorb in the UV region at 280 nm.

Evaluation of film thickness can be done by micrometer measurement if sample thickness is higher than 50 μm (meaning that the error of measurement allowed is less than 2%); in the case of thinner samples, measurement of the optical density of IR peak at 2920 cm^{-1} is applied, and results are calculated from the equation:

$$e_{cm} = OD_{2920}/172$$

From Verdu´s studies reported in Chapter Three, it is evident that film thickness affects the changes recorded. The formation of CO groups versus film thickness was studied by Verdu, but the other changes studied are also thickness related, and therefore, film thickness should always be seen as a possible factor.

Caraculacu (56) applied UV spectrophotometry for monitoring the concentration of phenolic groups reacted with polymer.

6.6.2. IR spectrophotometry

The application of IR spectroscopy has proved to be so rewarding in PVC degradation studies that in recent years it has probably become the most useful technique. Usually IR spectrophotometry is known from qualitative applications in chemistry, but in PVC degradation studies, it is mainly used for quantitative determinations.

Let us discuss numerous studies in the order of increasing wavelength in the IR region. The part of the spectrum above 3000 cm^{-1} is not so frequently applied in these studies. Rabek (48), studying peroxidation of PVC by the presence of oxidation products of tetrahydrofuran left in trace quantities in PVC films, used absorption in a range of 3440 and 3400 cm^{-1} to monitor the concentration of OH and OOH groups, respectively. Bellenger (71), interested in carbonyl group concentration, used absorption at 4400 cm^{-1} as an internal standard with which the absorption of the carbonyl group was compared. This differs from works by Cooray (72) and Verdu (70), who found that the band at 2920 cm^{-1} can serve as an internal standard since it is stable throughout the experiment, a logical consequence of its being the CH_2 band. Verdu (70) also used this band for sample thickness measurement. Amer (36) constructed equipment to study the HCl catalytic effect on the PVC dehydrochlorination rate in which HCl was continuously measured by IR gas cell at 2963 cm^{-1}.

Absorption at around 1700 cm^{-1} is probably the most frequently applied technique so far as IR spectroscopy is concerned, because many authors (47,54,67,70,72,73) found it to be appropriate for carbonyl group determination during the thermooxidation process. The range of values is quite broad, from 1715 to 1730, with a value of 580 l/mol cm given by Braun, (54) and 200-600 by Verdu (70). Rabek (48) used this part of the spectrum to determine the concentration of α-hydroxy-tetrahydrofuran and butyrolactone, measuring absorption at 1735 and 1789 cm^{-1}, respectively. Mano (74) presented data on PVC-ketone interaction, which according to him, are evident from absorption measured between 1719-1708 cm^{-1}. Then Mukherjee (42) found the band at 1635 cm^{-1} appropriate for monitoring the polyene growth rate.

Absorption between 1600 and 1200 cm^{-1} and below is also of interest for our studies, since it can be used for measuring the tacticity index (A_{1428}/A_{1434}), as suggested by Millán (43,47). Koenig (75) shows how one can apply Fourier transform IR to determine dioctylphthalate concentration from measurements in the 1600-600 cm^{-1} range. The same author (75) observed that bands at 1427, 1336, 1526, 1104, 956, 638 and 604 cm^{-1} are all related to internal order (tacticity). Also, measuring of inorganic and organic volatile products of PVC decomposition is possible in this range (75). Abbås (50) calculated the concentration of methyl branches from the following equation:

$$CH_3/1000C = KA_{1378}/dt$$

where:

A_{1378} - absorbance of sample at 1378 cm^{-1},

d - sample density,

t - sample thickness,

K - constant equal to 9.9.

The data calculated from IR measurement give correct values if the appropriate extinction coefficient is applied.

Finally, the other very important area of the IR spectrum in the range of 700-600 cm^{-1} was discussed by authors involved in studies of conformation and crystallinity (76,77). Sample form is of high importance, as can be found from Carrega's paper (78). The tacticity index, evaluated from measurements of cast film, was about 50% higher than that measured in KBr disc form. Actually, in the above studies all possible forms of specimens, i.e., KBr disc, film, solution and nujol mull were applied.

Summarizing the above review of IR methods, one may certainly suggest that the values obtained are usually valid in the range of a particular experiment since there is still little theoretical explanation for the method applied on the empirical basis. This is why crystallinity is frequently attributed to conformation changes, and vice versa.

6.6.3. Nuclear magnetic resonance spectroscopy

The main application of NMR studies is related to long branches, products of phenolysis reaction and end-group analysis. Starnes (63,64,79) gave a detailed description of NMR spectroscopy potential that has already been discussed throughout this book, and Abbås (80) made measurements using Starnes´ method. Caraculacu (50) and Hjertberg (58) applied NMR for determination of phenol incorporated in the place of labile chlorine, while Hjertberg was also successful in the identification of terminal groups in the PVC chain by using this technique. Wirth (81) described the more unusual application of NMR to study the interaction of different organotin chlorides. Ando (82) applied ^{1}H-NMR in conformation studies of meso and racemic diads in PVC.

Starnes (63,64,79), who experimented with various reducing agents, has made excellent use of the full capabilities of this method by utilizing Bu_3SnD as a reducing moiety. Monodeuteration of ^{13}C atom converts its proton-decoupled NMR signal from a singlet into a triplet, due to which this signal and those from neighboring carbon undergo upfield shifts. Based on these observations, one is able to identify the original points at which chlorine was bound. Since the structure of the chain backbone can be studied from the same measurement, the structure of the original polymer can therefore be deduced from the data collected according to Starnes´ method.

6.6.4. Electron spin resonance spectroscopy

ESR has an even more specific application since it is used mainly for photolytic studies of PVC degradation. Yang´s data (83) are especially valuable since they helped to clarify the stages of photolytic degradation and to identify the radical formed in the process of photodegradation. Rabek (48) employed ESR to study tetrahydrofuran participation in photooxidation, while many of his studies (discussed in Chapter Three) also helped to delineate the mechanism of photolysis. Mori (73) found an application for the method in studies of PVC thermal stabilizer participation in the PVC photolytic process.

Samples for measurements are either in the form of foil, as in Mori´s studies (73), or a solution (5-15% w/v) in tetrahydrofuran, dioxane or tetrahydropyran. The temperature at which measurement is done seems to be crucial for the type of radical detected, as discussed in detail by Young (83), but still one observes that some studies are done at room temperature (73).

6.6.5. Raman spectroscopy

It is surprising to notice that Raman spectroscopy , which theoretically is well suited for PVC studies, is not as popular as could be expected. It is already evident (84) that this method can be very useful in structural studies since it has one advantage over NMR and IR: The sample requires no special preparation as the powder form can be easily analyzed. Unlike in the other two methods, the sample retains its original structure. Data from Robinson's work (84) show that peaks at 647 and 634 cm^{-1} should be assigned to syndiotactic and isotactic structures, respectively. In studies of plasticized PVC it was shown that absorption at 647 cm^{-1} considerably decreases with an increase in plasticizer content, which confirms the right peak assignment.

It is hoped that this technique will be more fully explored in the future since there are many questions to be answered so far as chain conformation and configuration and PVC crystallinity are concerned.

6.7. Other methods used in PVC degradation studies

X-ray diffraction

This method, successfully applied in many polymer studies, has also been used for PVC investigation. Typically, a wide-angle X-ray diffraction technique is applied and a PVC sample in powder form used (76,77,85). X-rays are generated mainly from a $CuK\alpha$ source with a nickel filter. Occasionally, $MoK\alpha$ radiation is used, which allows studies at higher penetration. X-ray diffractometry gives comparable results; the only problem is that no one knows how to interpret them quantitatively. The typical scan is given by Fig. 6.12.

The crystallinity index is calculated from the following equation:

$$X = \phi_c/(\phi_a + \phi_c)$$

256

therefore, knowledge of the surface area under the curve for the amorphous sample and the shape of that curve are essential for the value obtained from calculations. Unfortunately, it has not yet been possible to establish a standard of crystalline material or to discover the method of amorphous sample preparation. Thus, results are disputable and comparable only in a particular experiment. X-ray studies have little relevance in PVC investigations.

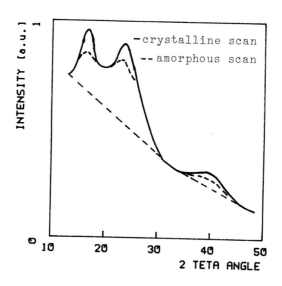

Fig.6.12. Wide angle X-ray diffraction pattern in comparison to amorphous pattern. (Modified from Ref. 76.)

Gas chromatography

Several authors have employed gas chromatography and mass spectroscopy coupled with gas chromatography for studies of the products of combustion or pyrolysis of PVC samples. Pyrolysis is conducted in a pyroprobe under helium or in the presence of oxygen at a temperature of up to 1273K, though usually only up to 873K (35,86,87). The heating rate of the pyroprobe is regulated in a range of 5-40 K/min. For the analysis of gas evolved, Lattimer (86) used a Varian 3700 GC coupled with a Finnigan MAT 311A mass spectrometer and Finnigan Incos 2400 data system, which is the best instrument for these studies. The spectrum of

pyrolytic gases analyzed by instrument is compared with an
existing library of individual compounds that assists
identification. In the case of Liebman´s studies (65), between
the pyroprobe and the GC unit, a six-port sampling valve was
interfaced, which allowed repetitive sampling every two minutes.
Ballistreri (87) measured the products of pyrolysis directly in a
mass spectrometer. Lattimer´s method (86) should be viewed as
the standard technique for these studies.

Gas chromatography was also applied by Bowmer (88) for studies
of short chain branches. PVC samples, being formerly
γ-irradiated in ampoules, were then heated in the injection
system of a gas chromatograph at 323-473K for 5-30 min and
crushed. Volatile products introduced into the carrier gas
stream were separated in a GC column.

Rheological properties of PVC in solution

This is the alternative method of long branch estimation, which
was used before chemical modification was applied. Initially,
results from rheological studies were misinterpreted, but the
uncertain points seem to be clarified now. According to Abbås
(89), the best results can be obtained using Zimm´s-Stockmayer´s
theory (89), which gives the relationship for trifunctional
branch points as follows:

$$\langle g_3 \rangle = \frac{6}{n_w}[0.5(\frac{2 + \bar{n}_w}{n_w})^{0.5} \ln\{\frac{(2 + \bar{n}_w)^{0.5} + \bar{n}_w^{0.5}}{(2 + \bar{n}_w)^{0.5} - \bar{n}_w^{0.5}}\} - 1]$$

where:

$\bar{n}_w$ - the weight average number of branch points per molecule,

$\langle g_3 \rangle^{0.5} = [\eta_{br}]/[\eta_l]$

$[\eta_{br}]$ - the intrinsic viscosity for branched polymer,

$[\eta_l]$ - the intrinsic viscosity for linear polymer.

Due to Abbås´ broad studies and analysis, the long branches can
be estimated by this method in their usual concentrations in PVC,
with an error rate up to 40%, compared with an error rate below
10% for the method of Bu_3SnH reduction discussed earlier.

Gel contents determination

Hjertberg (90) determined gel contents by gravimetric analysis of the insoluble material (3 hrs dissolution in THF at 393K) that did not pass a filter of 0.5 nm pore size.

Iván (91) concentrated his studies on soluble fraction, monitoring specific absorbance of the solution during degradation. The decrease in absorbance is attributed to the decrease in solubility due to gel formation. Comparison between both methods of gel determination from viscosity measurement and photometric method is given.

Chlorine determination

This analytical technique mainly serves purposes related to PVC stabilization, but it can also be applied to PVC degradation studies with the stabilizer playing the role of HCl acceptor. Several authors determined Cl^- ion concentration coulometrically (92-94). In this method, Ag^+ ion is continuously formed by anodic generation, which is automatically controlled.

Other methods

Ceccorulli (95) conducted calorimetric measurements by means of a differential scanning calorimeter to monitor glass transition changes and determine the content of undissolved stearic acid in mixtures with PVC. Hay (96) reported studies on similar equipment, but focussed on investigation of the nature of crystalline properties of PVC. The same type of studies are discussed by Carrega (77). Gilbert applied a liquid scintillation counter in order to explain PVC-epoxide interaction (52).

REFERENCES

1. D.D. Perin, W.L.F. Armarego and D.R. Perrin, in **Purification of Laboratory Chemicals,** Pergamon Press, Oxford, 1980.
2. Z.V. Popova and J. Janovski, **Zhur. Prik. Kchim.,** 33(1960)186.
3. Z.V. Popova and J. Janovski, **Zhur. Prik. Kchim.,** 34(1961)1324.
4. **ISO**/R-305/1963.

5. K.M. Ball and P.L. Kolker, **Brit. Plast.**, 42(1969).

6. K. Thinius, R. Schlimper and E. Kestner, **Plaste u. Kaut.**, 17(1970)804.

7. J. Novak, **Kunststoffe**, 51(1961)712.

8. P. Hedvig and W. Kibenyj, **Angew. Makromol. Chem.**, 7(1969)198.

9. A.Crosato-Arlandi, G. Palma, E. Peggion and G. Talamini, **J. Appl. Polym. Sci.**, 8(1964)747.

10. L.S. Troickaya and V.M. Miakov, **Vyssokomol. Soed.**, 9(1967)2119.

11. M. Lisý, **Chemické Zvesti**, 19(1965)84.

12. W. Reicherdt, Z. Wolkober and H. Krause, **Plaste u. Kaut.**, 13(1966)454.

13. **ASTM** D 793-49 (1965).

14. L.M. Wartman, **Ind. Eng. Chem.**, 47(1955)1013.

15. A. Guyot and J.P. Benevise, **J. Appl. Polym. Sci.**, 6(1962)98.

16. T.Q. Pham and P. Roux, **Chim. Anal.**, 48(1968)448.

17. DIN 53381/3 (1964).

18. A. Cittadini and R. Palillo, **Chim. Ind.**, 41(1959)980.

19. W.C. Geddes, **Eur. Polym. J.**, 3(1967)267.

20. M. Laczkó, **Muanyag Gumi**, 7(1970)15.

21. J. Wang and W. Tsai, **J. Chinese Inst. Chem. Eng.**, 10(1979)97.

22. M. Ohta and M. Imoto, **J. Chem. Soc. Japan**, 54(1951)470.

23. E.J. Arlman, **J. Polym. Sci.**, 12(1954)543.

24. B.Baum and L.H. Wartman, **J. Polym. Sci.**, 28(1958)537.

25. G. Talamini and G. Pezzin, **Makromol. Chem.**, 39(1960)26.

26. D. Braun and M. Thallmaier, **Kunststoffe**, 56(1966)80.

27. G. Schram, **Kunststoffe**, 58(1968)697.

28. A. Guyot and M. Bert, **J. Appl. Polym. Sci.**, 17(1973)753.

29. K.B. Abbås and E.S. Sörvik, **J. Appl. Polym. Sci.**, 17(1973)3567.

30. M. Farago and G. Garepian, **SPE-J.**, 26(1970)44.

31. D. Furnica and J.A. Schneider, **Makromol. Chem.**, 66(1967)102.

32. V.P. Gupta and L.E. Pierre, **J. Polym. Sci.**, 8(1970)37.

33. A.A. Caraculacu, E.C. Bezdadea and G. Istrate, **J. Polym. Sci.**, Part A-1, 8(1970)1239.

34. K. Figge and W. Findeiss, **Angew. Makromol. Chem.**, 47(1975)141.

35. S.A. Liebman, D.H. Ahlstrom and C.R. Foltz, **J. Polym. Sci.**,

Polym. Chem. Ed., 16(1978)3139.

36. A.R. Amer and J.S. Shapiro, J. Macromol. Sci.-Chem., A14(1980)185.

37. K.W. Egger, J. Amer. Chem. Soc., 91(1969)2867.

38. Z. Vymazal, E. Czakó, B. Meissner and J. Štěpek, J. Appl. Polym. Sci., 18(1974)2861.

39. J. Wypych, unpublished data.

40. ASTM D 1925-70.

41. Additives for PVC processing. Colorimetric evaluation of stability tests, Ciba-Geigy, Technical Bulletin.

42. A.K. Mukherjee and A. Gupta, J. Macromol.Sci.-Chem., 16(1981)1161.

43. J. Millán, G. Martinez and C. Mijangos, J. Polym. Sci., Polym. Chem. Ed., 18(1980)505.

44. W.R. Moore and R.J. Hutchinson, Nature, 200(1963)1095.

45. I.K. Varma and K.K. Sharma, Angew. Makromol. Chem., 79(1979)147.

46. G. Martinez and J. Millán, Angew. Makromol. Chem., 75(1979)215.

47. S. Matsumoto, H. Oshima and Y. Hasuda, J. Polym. Sci., Polym. Chem. Ed., 22(1984)869.

48. J.F. Rabek, T.A. Skowroński and B. Rånby, Polymer, 21(1980)226.

49. A. Wirsèn and P. Flodin, J. Appl. Polym. Sci., 22(1978)3039.

50. K.B. Abbås and E.M. Sörvik, J. Appl. Polym. Sci., 19(1975)2991.

51. E.E. Drott and R.A. Mendelson, J. Polym. Sci., A-2, 8(1970)1361.

52. J. Gilbert and J.R. Startin, Eur. Polym. J., 16(1980)73.

53. J. Gilbert, M.J. Shepherd, J.R. Startin and M.A. Wallwork, J. Chromatog., 237(1982)249.

54. D.Braun, A.Michel and D. Sonderhof, Eur. Polym. J., 17(1981)49.

55. A. Michel, E. Castaneda and A.Guyot, J. Macromol. Sci.-Chem., 12(1978)227.

56. A. Caraculacu, J. Macromol. Sci.-Chem. 12(1978)307.

57. E.C. Buruiană, A. Airinei, G. Robilă and A. Caraculacu, Polym. Bull., 3(1980)267.

58. T. Hjertberg and E.M. Sörvik, J. Macromol. Sci.-Chem.,

17(1982)983.

59. T. Morikawa, **Kagaku to Kogyo**, 41(1967)169.

60. J. Boissel, **J. Appl. Polym. Sci.**, 21(1977)855.

61. G. Scott, M. Tahan and J. Vyvoda, **Eur. Polym. J.**, 14(1978)377.

62. A. Guyot, M. Bert, P. Burille, M.-F. Llauro and A. Michel, **Pure Appl. Chem.**, 53(1981)401.

63. W.H. Starnes, **Dev. Polym. Deg.**, 3(1981)135.

64. M. Carrega, C. Bonnebat and G. Zednik, **Anal. Chem.** 42(1970)1807.

65. W.H. Starnes, R.l. Hartless, F.C. Schilling and F.A. Bovey, **ACS Polym. Prep.**, 18(1977)499.

66. D. Braun, **Pure Appl. Chem.**, 53(1981)549.

67. J.F. Rabek, B. Rånby, B. Östensson and P. Flodin, **J. Appl. Polym. Sci.**, 24(1979)2407.

68. Z. Vymazal, E. Czakó, K. Volka, J. Štěpek, R. Lukás, M. Kolinský and K. Bouchal, **Eur. Polym. J.**, 16(1980)151.

69. R.S. Davidson and R.R. Meek, **Polym. Photochem.**, 2(1982)1.

70. J. Verdu, **J. Macromol. Sci.-Chem.**, 12(1978)551.

71. V. Bellenger, J. Verdu, L.B. Carette, Z. Vymazalová and Z. Vymazal, **Polym. Deg. Stab.**, 4(1982)303.

72. B.B. Cooray and G. Scott, **Polym. Deg. Stab.**, 3(1980-81)127.

73. F. Mori, M. Koyama and Y. Oki, **Angew. Makromol. Chem.**, 68(1978)137.

74. E.B. Mano and E.E.C. Monteiro, **J. Polym. Sci., Polym. Let. Ed.**, 19(1981)155.

75. J.L. Koenig and M.K. Antoon, **Appl. Optics**, 17(1978)1374.

76. R.P. Chartoff, T.S.K. Lo, E.R. Harrel and R.J. Roe, **J. Macromol. Sci.-Phys.**, 20(1981)287.

77. R. Biais, C. Geny, C. Mordini and M. Carrega, **Brit. Polym. J.**, (1980)179.

78. M.E. Carrega, **Pure Appl. Chem.**, 49(1977)569.

79. W.H. Starnes, F.C. Schilling, I.M. Plitz, R.E. Cais and F.A. Bovey, **Polym. Bull.**, 4(1981)555.

80. K.B. Abbås, **Pure Appl. Chem.**, 53(1981)411.

81. H.O. Wirth, H.A. Miller and W. Wehner, **J. Vinyl Technol.**, 1(1979)51.

82. I. Ando, **Makromol. Chem.**, 179(1978)2663.

83. N.L. Yang, J. Liutkus and H. Haubenstock, **ACS Symp. Ser.**,

182(1980)35.

84. M.E.R. Robinson, D.I. Bower and W.F. Maddams, **Polymer**, 19(1978)773.

85. S.J. Guerrero, D. Meader and A. Keller, **J. Macromol. Sci.-Phys.**, 20(1981)185.

86. R.P. Lattimer and W.J. Kroenke, **J. Appl. Polym. Sci.**, 27(1982)1355.

87. A. Ballistreri, S. Foti, G. Montaudo and E. Scamporrino, **J. Polym. Sci., Polym. Chem. Ed.**, 18(1980)1147.

88. T.N. Bowmer, S.Y. Ho, J.H. O´Donnell, G.S. Park and M. Saleem, **Eur. Polym. J.**, 18(1982)61.

89. B.H. Zimm and W.H. Stockmayer, **J. Chem. Phys.**, 17(1949)1301.

90. T. Hjertberg and E.M. Sörvik, **J. Appl. Polym. Sci.**, 22(1978)2415.

91. B. Iván, B. Turcsányj, T.T. Nagy, T. Kelen and F. Tüdös, **Polymer**, 19(1978)351.

92. J. Štěpek, Z. Vymazal and E. Czakó, **J. Macromol. Sci.-Chem.**, 12(1978)401.

93. T.T. Nagy, T. Kelen, B. Turcsányj and F. Tüdös, **Polym. Bull.**, 2(1980)749.

94. M.W. Mackenzie, H.A. Willis, R.C. Owen and A. Michel, **Eur. Polym. J.**, 19(1983)511.

95. G. Ceccorulli, M. Pizzoli, M. Scandola, G. Pezzin and G. Crose, **J. Macromol. Sci.-Phys.**, 20(1981)519.

96. J.N. Hay, F. Biddlestone and N. Walker, **Polymer**, 21(1980)985.

INDEX